Mechanisms in Transcriptional Regulation

Albert J. Courey

University of California, Los Angeles

Blackwell
Publishing

© 2008 by Blackwell Publishing

BLACKWELL PUBLISHING
350 Main Street, Malden, MA 02148-5020, USA
9600 Garsington Road, Oxford OX4 2DQ, UK
550 Swanston Street, Carlton, Victoria 3053, Australia

The right of Albert J. Courey to be identified as the Author of this Work has been
asserted in accordance with the UK Copyright, Designs, and Patents Act 1988.

First published 2008 by Blackwell Publishing Ltd

1   2008

Library of Congress Cataloging-in-Publication Data

Courey, Albert J.
  Mechanisms in transcriptional regulation / Albert J. Courey.
    p.; cm.
  Includes bibliographical references and index.
  ISBN-13: 978-1-4051-0370-1 (pbk. : alk.paper)
  ISBN-10: 1-4051-0370-1 (pbk. : alk.paper)
  1. Genetic transcription – Regulation.   2. Transcription factors.   I. Title.
  [DNLM: 1. Eukaryotic Cells – physiology.   2. Transcription, Genetic.   3. Bacteria –
genetics. QH450.2 C859m 2008]

  QH450.2C68 2008
  572.8′845 – dc22   2007038526

A catalogue record for this title is available from the British Library.

Set in 10/12.5 Meridien
by Graphicraft Limited, Hong Kong
Printed and bound in Singapore
by Fabulous Printers Pte Ltd

The publisher's policy is to use permanent paper from mills that operate a sustainable
forestry policy, and which has been manufactured from pulp processed using acid-free and
elementary chlorine-free practices. Furthermore, the publisher ensures that the text paper
and cover board used have met acceptable environmental accreditation standards.

For further information on
Blackwell Publishing, visit our website:
www.blackwellpublishing.com

Cover illustration: The Polycomb chromodomain (Protein Data Bank id 1PDQ) complexed
to a portion of the histone H3 N-terminal tail containing trimethyl-lysine 27. This structure
was first described in: Fischle, W., Wang, Y., Jacobs, S. A., Allis, C. D. and Khorasanizadeh,
S. (2003) Molecular basis for the discrimination of repressive methyl-lysine marks in
histone H3 by Polycomb and HP1 chromodomains. *Genes Dev*, **17**, 1870–1881.

Short contents

Color plate section between pp. 82 and 83

Dedicated to Jody, Sibyl, and Ethan for their unconditional love and support

Full contents

Color plate section between pp. 82 and 83

List of boxes

Preface

The impetus for this book is a course that I teach at UCLA – "Mechanisms of Eukaryotic Transcriptional Regulation". While this course focuses on eukaryotic transcription, almost all of the students take it just after completing a course entitled "Mechanisms of Prokaryotic Transcriptional Regulation" taught by my colleague Jay Gralla. In Jay's course, the students are introduced to a number of ideas in bacterial transcription that greatly aid an understanding of eukaryotic transcription. Thus, although this book focuses on eukaryotes, I have included key ideas and examples from bacterial transcription.

Like my course, this book is intended for both upper division undergraduates and graduate students in the molecular life sciences. It should also be useful to more senior scientists who find that their research has, either by accident or by design, entered the realm of transcription, and who therefore require a basic introduction to the field.

This book is by no means a comprehensive account of transcription, since a complete treatment of this large and rapidly expanding field would require multiple volumes. Instead, it discusses a few topics that are essential to an appreciation of the field and illustrates these topics with a number of carefully selected examples. These topics include the workings of the basal transcriptional machinery (Chapters 2 and 3), mechanisms of activation (Chapter 4), the role of chromatin in eukaryotic transcriptional control including mechanisms of epigenetic regulation (Chapters 5 and 6), and mechanisms of combinatorial control (Chapter 7).

While this book emphasizes unifying ideas, it is only through an appreciation of the details that underlie these ideas that one gains a deep understanding of them. Therefore, I have attempted to include enough examples to help readers feel at home with the concepts and the experimental basis behind them. My goal is to equip readers with the ability to approach the primary scientific literature with a critical mind.

While most of the conclusions presented in this book are broadly accepted, transcription is a rapidly evolving field being actively pursued in thousands of research labs around the world. In an effort to capture some of the current excitement in the field, I have chosen to cover certain topics that are not yet completely settled. Rather than attempting to present all sides of every issue, I have, in some cases, chosen to take a particular stance that I believe is supported by the currently available evidence. Thus, some of the conclusions presented in this book will no doubt require future revision.

To increase accessibility, each chapter includes essential background information set off from the main text. These boxes introduce relevant experimental approaches (e.g., protein chromatography, the use of chemical probes, chromatin immunoprecipitation assays, genetic suppression) and relevant topics in biology and biochemistry (e.g., protein structure visualization, cooperativity, *Drosophila* segmentation). In addition, a few essential terms that are not defined in the main text are explained in the margins, and a comprehensive list of definitions is provided in the form of a glossary at the end of the book (terms highlighted in color in the text).

Since it is just as important to understand what we do not know as it is to understand what we do know, explicit discussions of some of the most important open questions is presented in side boxes. Furthermore, each chapter also includes a set of problems, many of which bring up complex unresolved issues and are intended to stimulate discussion. Possible answers to these problems are provided at the end of the book.

Finally, each chapter contains a list of suggested further reading grouped and ordered to relate to the way that the material is presented in the text. These are not meant to be comprehensive bibliographies, but instead include a few articles mainly from the primary research literature that have influenced my thinking and that have withstood the test of time. In generating these reading lists, I have left out many equally important and worthy papers, and for this, I apologize to my colleagues in the scientific community.

Albert Courey

Acknowledgments

I would like to acknowledge my colleagues at UCLA, including the members of my research group and the Department of Chemistry and Biochemistry, for their support as I devoted long hours to this project. I especially thank Jay Gralla, with whom I began co-teaching the Mechanisms of Transcriptional Regulation course series at UCLA 17 years ago. It is thanks to Jay, whose research delves into mechanisms of both bacterial and eukaryotic transcription, that I came to realize that bacteria and eukaryotes are not such different beasts after all, and that an appreciation of one is essential to an understanding of the other. The comments, questions, and responses from our many students over the years have helped me to improve my presentation of the beautiful intricacies of transcriptional control.

My affiliation with the UCLA Interdepartmental Graduate Program in Gene Regulation has played a central role in shaping my thinking about the subject of this book. The weekly journal club associated with this program is a forum for penetrating discussion of the latest developments in the field of transcription. My frequent exchanges with such colleagues as Arnold Berk, Michael Carey, Michael Grunstein, Siavash Kurdistani, and Stephen Smale have been extremely stimulating.

This book has benefited greatly from the insightful comments of James Kadonaga (UC San Diego), Songtao Jia (Columbia University), and Stephen Small (New York University). Their thought-provoking observations and questions have been invaluable in my efforts to improve the readability and focus of the text.

I am also grateful to Blackwell Publishing for shepherding this book through to completion: Nancy Whilton visited me year after year before finally convincing me to take on the project; Elizabeth Frank and Haze Humbert showed great patience when I missed deadlines; Steve Weaver and Karen Chambers continued to support the project after those who initiated it at Blackwell moved on to other things; Jane Andrew paid great attention to detail during the proofreading process. Special thanks are also due to Chris Lear and his colleagues at J&L Composition for their patience in repeatedly revising the illustrations.

Finally, my deepest thanks go to James C. Wang (Harvard University) and Robert Tjian (UC Berkeley) for teaching me that one can only appreciate the big picture after a thorough consideration of the details.

1

The vocabulary of transcription

Key concepts

- Mechanisms of transcription and its regulation are conserved across the domains of life
- The basal machinery, which catalyzes transcription, consists of core RNA polymerases and accessory factors
- The basal machinery is controlled by a regulatory machinery consisting of activators, repressors, coactivators, and corepressors

1.1 INTRODUCTION

Proteins control nearly everything that happens in living organisms – they decode genetic information; they control cell shape, cell movement, and segregation of the chromosomes during cell division; they determine the localization of every cellular constituent; and catalyze every essential metabolic reaction. So if we are to understand how organisms work, we need to understand the processes that control when and where each protein is synthesized.

The production of a protein is a multistep process beginning with the synthesis of RNA (transcription), the processing of this RNA, and its transport to the ribosomes where it serves as a template for the synthesis of polypeptides (translation). These polypeptides must then be correctly modified and folded to form mature proteins.

Although the pathway from a gene to a mature protein can be regulated at any step, the most efficient way to regulate gene expression is to modulate the first step in the pathway since this avoids the costly synthesis of unneeded RNA. Organisms therefore produce an immense set of catalytic and regulatory factors dedicated to the task of transcribing genes in an intricately controlled manner. This chapter will introduce the components of this machinery and discuss the extent to which these components have been conserved during the course of evolution. At the same time, readers will become acquainted with some of the vocabulary needed to appreciate the exploration of the transcriptional machinery presented in Chapters 2 through 7.

1.2 THE VOCABULARY OF TRANSCRIPTION

Readers of this book should have had some previous exposure to most of the terms given in bold face in this section, and to basic ideas about proteins and nucleic acids through a college level course in biochemistry or molecular biology. Other important terms defined in this section, to which readers may or may not have previous exposure, will be given in italics.

1.2.1 RNA biogenesis

Transcription, the first step in *RNA biogenesis*, is the DNA template-dependent synthesis of RNA, a process that is catalyzed by **RNA polymerases**. Special DNA sequences termed **promoters**, which are found near the beginning of each transcribed region, direct these enzymes to initiate transcription. *Transcriptional initiation* is followed by *transcriptional elongation*, during which polymerase moves along the DNA, catalyzing the template-directed joining of nucleotides via **phosphoester linkages** to form a full-length *primary transcript* that is complementary in sequence to one of the DNA strands (the "template strand"). Elongation is, in turn, followed by *transcriptional termination*, in which the polymerase encounters a termination signal triggering the release of both the transcript and the DNA from the polymerase.

The job of most RNA molecules is to direct protein synthesis by serving as a template (**mRNA**) or as components of a catalytic machine for decoding this template (**tRNA** and **rRNA**). The production of RNA molecules ready to participate in protein synthesis requires numerous modifications to the primary transcripts. rRNA and tRNA modification occurs in all organisms, while mRNA modification (mRNA processing) is largely restricted to eukaryotes. Eukaryotic mRNA processing includes *5′ capping* to produce the mature 5′ end of the transcript, *splicing* to remove introns, and *cleavage and polyadenylation* to produce the mature 3′ end. These processing events begin before synthesis of the primary transcript is complete (Figure 1.1).

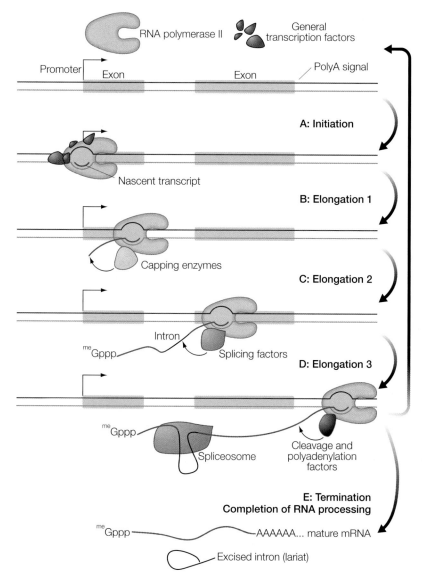

Figure 1.1 *RNA polymerase II is an mRNA factory.* RNA polymerase II (Pol II) interacts with a myriad of transcription factors and RNA processing factors to coordinate the process of mRNA biogenesis. (A) Promoter recognition and transcriptional initiation by Pol II requires general transcription factors. These proteins form a complex bound to the promoter, which then opens up the DNA exposing the template strand. Although it is not shown, after Pol II leaves the promoter and begins elongation, some of the basal factors remain behind at the promoter, others remain bound to the elongating Pol II, and others are released back into solution. (B–D) Early during the elongation phase, Pol II binds the enzymes that catalyze capping, which then direct the formation of a ^{me}Gppp cap in which 7-methylguanosine is attached to the 5′ end of the transcript via a 5′ to 5′ triphosphate linkage.

The RNA polymerase responsible for eukaryotic mRNA synthesis (RNA polymerase II) is sometimes described as an mRNA factory because it couples transcription to most of the processing steps required for the maturation of the mRNA (Chapter 3). By binding to capping, splicing, and cleavage and polyadenylation factors, RNA polymerase II ensures that appropriate sites in the primary transcript will be efficiently delivered to the processing machinery while transcription is still occurring (Figure 1.1).

1.2.2 The transcriptional machinery

The *transcriptional machinery* (or transcriptional apparatus) is the set of proteins required to carry out and regulate transcription. Some features of the transcriptional machinery are common to all organisms, while others are specific to one or two of the *three domains of life* (*bacteria, archaea,* and *eukaryotes*; Box 1.1).

The components of the transcriptional machinery are frequently divided into two categories: the *basal machinery*, which is directly responsible for promoter recognition and transcription, and the *regulatory machinery*, which controls the rate at which the basal machinery carries out its job in a gene-specific manner.

Basal machinery

The basal machinery in each organism includes one or more multisubunit *core RNA polymerases*, which are highly conserved in all cellular life (Chapter 2). Bacteria and archaea each contain a single core polymerase, while eukaryotes contain at least three core polymerases (RNA polymerase I, RNA polymerase II, and RNA polymerase III). Core RNA polymerases are able to catalyze DNA-dependent RNA synthesis, but they are not able to recognize promoters. In a test tube, they can be made to initiate transcription at gaps and nicks in double-stranded DNA, but this is a non-physiological process that does not normally occur in cells. Thus, the basal machinery also includes additional protein factors that

Figure 1.1 (*continued*) As elongation proceeds, Pol II binds splicing factors, which then trigger the formation at each intron of a spliceosome (a large ribonucleoprotein complex that catalyzes splicing). Pol II also binds the factors that catalyze cleavage of the transcript at the PolyA signal and the addition of several hundred adenylate residues to the newly formed 3' end (polyadenylation). (E) After Pol II passes the PolyA signal, termination occurs releasing the transcript and DNA template from Pol II. The splicing and formation of the mature 3' end may not be completed until after termination. The mature mRNA contains the 5' cap and 3' PolyA tail, which help to stabilize the RNA and are required for efficient export and translation of the mRNA. In the process of splicing together the exons, the introns are excised from the transcript and released as lariats.

Box 1.1

The domains of life

All organisms are either eukaryotes, which contain membrane-bound organelles (e.g., nuclei, mitochrondria, etc.), or prokaryotes, which lack membrane-bound organelles. Until about 30 years ago, it was taken for granted that these two types of organisms represented the only two main branches on the tree of life. It therefore came as a surprise when DNA sequence analysis led to the conclusion that there were two completely distinct types of prokaryotes (now termed bacteria and archaea), and that these two groups were as different from one another as either was from the eukaryotes. In 1977, these observations lead C.R. Woese to suggest the now generally accepted idea that there are three main branches on the tree of life, the bacteria, the eukaryotes, and the archaea (Figure B1.1).

Archaea were first discovered in very harsh environments such as thermal vents at the bottom of the ocean where the water temperature can exceed 100 °C. These environments resemble the conditions that existed on earth billions of years ago when cellular life was first evolving, leading to speculation that archaea (which means old) are a particularly ancient form of life. Many features of archaeal biochemistry distinguish this group of organisms from both bacteria and eukaryotes, perhaps explaining how these organisms can withstand such harsh conditions. The phospholipids in archaeal membranes, for example, have a unique structure that is different from the structure of bacterial and eukaryotic phospholipids.

Archaea appear to contain a blend of eukaryotic and bacterial features. Like bacteria, their genomes are organized into single circular chromosomes and their genes are frequently organized into operons. In contrast, the archaeal information-processing machinery (e.g., the DNA replication, transcriptional, and translational machinery) is, in many respects, more similar to the eukaryotic machinery than to the bacterial machinery (see Chapters 2 and 3 for ways in which the archaeal and eukaryotic basal transcriptional machinery are more similar to one another than they are to the bacterial basal machinery). The similarity between the information processing machinery in eukaryotes and archaea provides support for the idea that archaea and eukaryotes arose from a common ancestor that is distinct from the ancestor of modern bacteria (Figure B1.1).

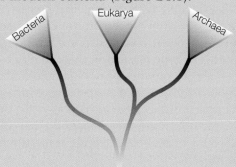

Progenote

Figure B1.1 *The three-branched tree of life.* The progenote is a hypothetical organism that is the ancestor of all currently existing life. The idea that archaea and eukarya arose from a common branch that is distinct from the bacterial branch is broadly although not universally accepted.

assist core RNA polymerases in promoter recognition and transcriptional initiation.

The factors required for promoter-specific transcription vary considerably depending on the core RNA polymerase. In the case of bacterial core RNA polymerase, promoter specificity requires a single factor termed a σ *factor* (Chapter 2). However, eukaryotes and archaea lack σ factors; instead their core RNA polymerases employ multiple *general transcription factors* for promoter-specific initiation (Chapter 3). While eukaryotic and archaeal general transcription factors are related to each other, they show no clear relationship to the bacterial σ factors. In contrast to the simplicity of the bacterial basal machinery, the eukaryotic basal machinery can be complex. For example, RNA polymerase II employs six general transcription factors containing more than 20 polypeptides to initiate transcription.

Regulatory machinery

The regulatory machinery in all organisms includes **activators**, which stimulate transcription by the basal machinery, and **repressors**, which inhibit transcription by the basal machinery (Chapter 4). Activators and repressors function in a gene-specific manner and thus are able to determine which genes are being actively transcribed in each particular cell type under any particular set of environmental conditions. Because this gene specificity depends on the ability of these activators and repressors to recognize and bind specific DNA sequences in the genes being regulated, activators and repressors are collectively termed *sequence-specific transcription factors*. Once bound to the DNA, these factors influence rates of transcription by a variety of mechanisms. In bacteria and archaea, and sometimes in eukaryotes, they make direct contacts with the basal machinery and either interfere with the initiation process in the case of repressors or stimulate the initiation process in the case of activators.

Because of their greater complexity, eukaryotes have the need for more intricate patterns of gene regulation than do other organisms. To accommodate this need, eukaryotes possess an additional group of regulatory factors, termed *coactivators* and *corepressors*, which are essentially absent from bacteria (and probably from archaea as well). Unlike sequence-specific transcription factors, these *coregulators* do not bind to DNA directly. Instead, they bind to sequence-specific transcription factors that are, in turn, bound to DNA.

Coactivators and corepressors influence rates of transcription in many ways. Like activators and repressors they sometimes directly contact the basal machinery to inhibit or stimulate a step in transcription (Chapter 4). In addition, many coregulators control transcription by catalyzing covalent or non-covalent changes in the structure of **chromatin**, which is the nucleoprotein complex into which almost all eukaryotic DNA is packaged (Chapter 5). Chromatin is approximately equal parts protein and DNA, and the major protein component is the **histones**. These histones are subject to a variety of post-translational modifications such as *acetylation, methylation,* and *phosphorylation*. Coactivators and corepressors frequently signal to the rest of the transcriptional machinery by catalyzing the addition or removal of these modifications.

Modifications in chromatin structure that determine the transcriptional states of individual genes are often stable enough to survive **mitosis**. Because they can be inherited from cell generation to cell generation as if they resulted from alternative **alleles** of a gene rather than alternative forms of chromatin, these states are called *epigenetic states* (Chapter 6). Epigenetic states may be as important as **genotype** and the environment in dictating **phenotype**.

1.2.3 Cis-elements

A diffusible component of the transcriptional machinery such as an activator is historically termed a trans-acting factor to denote its ability to act on chromosomes other than the one that encodes it. In contrast, the DNA segments with which the machinery interact are termed **cis-elements** since they usually only influence the transcription of the chromosome in which they are found. Cis-elements include promoters as well as the binding sites for activators and repressors.

Cis-regulatory modules

In bacteria and archaea, each gene is usually regulated by one or two sequence-specific transcription factors. In eukaryotes, however, genes are often regulated by many factors, each one acting through multiple sites. These sites are often arranged in clusters of a few hundred to about a thousand basepairs in length. This clustering of sites allows multiple sequence-specific transcription factors to interact with one another in a process termed *combinatorial control* (Chapter 7).

In bacteria and archaea, binding sites for activators and repressors are usually located upstream of the promoter and within about 60 bp of the transcriptional start site. In contrast, clusters of cis-elements in eukaryotes are commonly found both upstream and downstream of the promoter and can be located hundreds of thousands or even millions of basepairs away from the promoter. Clusters of cis-elements that act over long distances were initially termed **enhancers** if they functioned in activation or **silencers** if they functioned in repression. However, since many of these cis-element clusters interact with both activators and repressors, the more generic term *cis-regulatory module* (CRM) is often used.

The mechanisms by which CRMs send regulatory signals to distant promoters are still poorly understood. In many instances, however, this

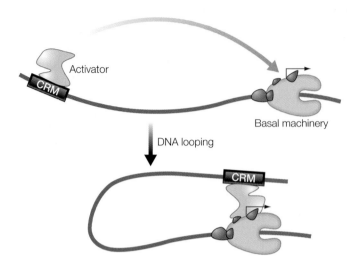

Figure 1.2 *Communication between cis-regulatory modules (CRMs) and promoters by DNA looping.* In eukaryotes, CRMs are often a long distance from the promoter. Communication between the CRM and the promoter may involve DNA looping, allowing direct contact between the regulatory factors (such as activators) bound to the CRM and the basal machinery bound to the promoter.

What controls the specificity of signaling between CRMs and core promoters?

For gene regulation to occur in an orderly manner, it is essential that each CRM only activates or represses an appropriate set of genes. Given that eukaryotic CRMs can be millions of basepairs away from their target promoters, proximity along the DNA does not seem to fully account for the specificity of CRM to promoter signaling. Some of the needed specificity may result from regions in the DNA known as boundary elements, which serve as barriers to CRM signaling. However, the physical nature of these boundaries and how they serve to limit CRM signaling are not understood.

signaling requires the formation of DNA loops bringing the CRMs into the proximity of the promoters allowing sequence-specific transcription factors and coregulators to contact the basal machinery (Figure 1.2).

1.3 EVOLUTIONARILY CONSERVED MECHANISMS OF TRANSCRIPTION

1.3.1 Conservation across the three domains of life

A few decades ago, the prevailing view was that bacterial and eukaryotic transcription were only distantly related to one another and that you could understand one without an appreciation of the other. This notion has not withstood the test of time. Biochemical and structural analysis of bacterial and eukaryotic RNA polymerases show that they are very closely related enzymes that catalyze transcription in very similar ways (Chapter 2). Furthermore, the mechanisms used to regulate bacterial transcription are conserved in eukaryotes (Chapter 4). Much has been learned about the catalysis and regulation of transcription using the very powerful approaches of bacterial genetics and biochemistry and these findings are included in this book, even though the central focus is eukaryotic transcription.

The third domain of life, the archaea, will be mentioned briefly in the chapters on the basal machinery (Chapters 2 and 3), but the regulation of transcription in archaea, a field that is still in its infancy, will not be discussed. It is worth pointing out, however, that the little we do know about the regulation of transcription in archaea reinforces the notion of conservation across the domains of life. This is because the archaeal transcriptional machinery seems to be a hybrid between the bacterial and eukaryotic machineries. As mentioned above, the archaeal basal machinery lacks σ factors and contains general transcription factors that are homologous to eukaryotic general factors. In contrast, archaeal sequence-specific transcription factors often have direct counterparts in bacteria and they seem to regulate transcription using the simpler set of regulatory mechanisms available in bacteria. Thus, archaea appear to use a bacterial-type regulatory machinery to regulate a eukaryotic-type basal transcriptional machinery, strongly suggesting conservation of mechanism across the domains of life.

1.3.2 Model eukaryotic organisms (and a plug for genetics)

Scientists are often driven to study transcription by their desire to understand how our own bodies decode genetic information leading to cell differentiation and the ability to respond to changes in the environment. It might seem that all such scientists should focus their studies on the human transcriptional machinery. This would be a mistake, however, since studies of model eukaryotic organisms have greatly enhanced our understanding of transcription in humans. The mechanisms of transcription are highly conserved throughout the eukaryotic domain, from protists, to fungi, to plants, to animals, and thus findings from studies of transcription in any eukaryote will illuminate the workings of the human transcriptional apparatus. The differences can be as telling as the similarities!

The development of genetics as a field of investigation during the last century was key to transforming biology from an observational to an experimental science. In classical genetic analysis, a population is exposed to a mutagen and then large numbers of individuals are screened for mutations affecting a pathway of interest. This allows the identification of genes encoding novel pathway components. Experimental manipulation of the mutants resulting from these screens often reveals the regulatory relationships between these components. In reverse genetic analysis, previously identified genes are altered and then placed back into an organism's genome. This makes it possible to see what happens to an organism upon changing a single amino acid in a component of the transcriptional apparatus or a single nucleotide in a cis-element. By combining these classical and reverse genetic approaches with biochemical and structural biological approaches, it is possible to build up a sophisticated view of the workings of the transcriptional apparatus.

Two model organisms that are often employed in studies of transcription because of the relative ease with which they can be subjected to genetic analysis are the budding yeast *Saccharomyces cerevisiae* and the fruit fly *Drosophila melanogaster*. As a result, these two organisms will be featured in many of the examples presented in this book. Both budding yeast and fruit flies are characterized by short life cycles, simple, well-characterized, and easily manipulated genomes, and powerful tools for the production and analysis of transgenic individuals. As a result, highly developed classical and reverse genetic tools are available to study the transcriptional machinery in both species greatly enhancing progress toward an understanding of this machinery.

1.4 WHAT'S COMING UP

A brief description of the remaining chapters is as follows.

Chapter 2: RNA polymerases and the transcription cycle

Chapters 2 and 3 present the nuts and bolts of the basal transcriptional machinery. This subject is of interest in its own right to anyone interested in such issues as fidelity in the transmission of biological information and the relationship between protein structure and function (Box 1.2). In addition, since the basal machinery is the target of the regulatory machinery, a sophisticated understanding of transcriptional regulation requires an understanding of the basal machinery.

Chapter 2 discusses the evolution of RNA polymerases and introduces the mechanisms of transcriptional initiation, elongation, and termination. The mechanisms by which RNA polymerases catalyze the elongation of RNA strands are highly conserved across all three domains of life, and this chapter will therefore draw on studies carried out in both bacteria and eukaryotes to illuminate the elongation cycle. In contrast, the machinery used for initiation has diverged significantly between bacteria and eukaryotes, and the discussion of initiation in Chapter 2 will focus on the process as it occurs in bacteria. However, the principles to be outlined are highly relevant to eukaryotic initiation, a subject that will be discussed in detail in Chapter 3.

Box 1.2

Looking at macromolecular structures

Macromolecules are microscopic machines with dimensions in the nanometer range. Just as it is impossible to understand how a macroscopic machine (e.g., an automobile motor) works without knowing how the parts are shaped and how they all fit together, one cannot hope to achieve an in-depth understanding of macromolecular function without an appreciation of macromolecular structure. Throughout this book, you will therefore be presented with various views of high resolution protein and nucleic acid structures to help you appreciate the enzymatic and regulatory mechanisms that govern transcription.

The major tools available to biochemists for determining macromolecular structure are X-ray crystallography (X-ray diffraction) and nuclear magnetic resonance (NMR) spectroscopy. In addition, these tools are sometimes supplemented with model building based, for example, on information from chemical probing experiments about where various macromolecules are thought to contact one another. The structures determined by these approaches are of varying quality, in part because macromolecules exhibit large amounts of disorder (thermal motion), which can lead to the erroneous interpretation of diffraction or NMR data. In addition, the very act of crystallizing a molecule can distort its structure due to crystal packing artifacts. Thus, it is best if structures have been validated by functional experiments that attest to their reasonableness. For example, the RNA polymerase structures determined by X-ray crystallography suggest the presence of one or more magnesium ions in the active site. This interpretation is validated by a number of enzymological and chemical probing experiments showing that metal ions are directly involved in catalysis by RNA polymerase.

Whenever a structural biologist publishes a macromolecular structure, he or she is obligated to deposit that structure into a searchable, publicly accessible, database termed the Protein Data Bank (http://www.rcsb.org). Each structure is deposited as a coordinate file (sometimes termed a PDB file), which lists the three-dimensional coordinates for each atom in the structure. Each coordinate file has a unique four-character PDB identification code, which can be used to look up a structure in the database, and which is often provided in journal articles when the structures are first published. Coordinate files for macromolecules of interest can also be located using the search engine on the Protein Data Bank web site. Molecular graphics programs can then be used to convert these PDB files into graphic representations of the structures, which is how the structures shown in this book were generated.

While the static pictures in this book are a good way to begin learning about structures, the best way to get a feel for a particular structure is to display it on your own computer using molecular graphics software so that you can render it in various formats, rotate it (just by dragging your mouse), and zoom in on features of interest. Powerful molecular graphics programs that allow you to do just this are available on the internet. Three such programs are listed below:

►

- **Protein Explorer** (http://proteinexplorer.org). This free program, developed by Eric Martz, has a powerful menu-driven interface and lots of online help. It runs through web browsers but is not compatible with all operating systems. Protein Explorer was used to create Plate B1.1.
- **FirstGlance** (http://firstglance.jmol.org). This free program, also developed by Eric Martz, is probably the easiest way to look at macromolecules. It is less powerful than Protein Explorer, but works through almost any modern web browser.
- **PyMOL** (http://pymol.sourceforge.net/). This program, developed by Warren DeLano, produces publication quality molecular graphics, but has a steeper learning curve than the above programs. It was used to create all the molecular graphics images in the color plate section, with the exception of Plate B1.1. PyMOL can be used for free by full-time students and their teachers. A subscription fee is required for academic researchers, industrial users, and independent contractors.

Molecular graphics programs provide a wide variety of display modes. These include:

1 The wireframe view (Plate B1.1B). Each bond is represented by a line of the appropriate length. The atoms themselves are not explicitly shown but are located at the vertices of the bonds. This is a good view for seeing the covalent structure of a macromolecule.
2 The spacefilling view (Plate B1.1E). Each atom is shown as a sphere, the radius of which is proportional to the atom's Van der Waals radius. This view gives you a good idea of the overall shape of the molecule and of the packing between residues and ligands.
3 The surface view (Plate B1.1A). This view reveals the solvent accessible surface of a molecule. It is the area that can be contacted by a probe (usually a sphere with the radius of a water molecule) as it is rolled over the macromolecule.
4 The backbone trace (Plate B1.1C). This is a smooth line that connects the α-carbons of a polypeptide chain, or the phosphorus atoms of a nucleic acid chain. Sometimes one or more side chains are added to the view. It is useful for seeing the path of the backbone and for seeing side chain interactions without the confusion that can be caused by all the backbone atoms.
5 The cartoon view (Plate B1.1D). This is a version of the backbone trace stylized so as to highlight protein secondary structure (α-helices and β-strands).

The use of color in these representations is often very important in allowing the viewer to identify important features, such as polypeptide chains, domains, residues, or individual atoms. In the frequently used CPK coloring scheme (named for the three scientists, Corey, Pauling, and Kultin, who devised it), carbon atoms are shown in gray, oxygen in red, nitrogen in blue, and sulfur in yellow.

Since it is just as important to understand what we do not know as it is to understand what we do know, explicit discussions of some of the most important open questions are presented in side boxes.

Chapter 3: The eukaryotic basal machinery

The primary focus of this chapter is the general transcription factors that assist eukaryotic RNA polymerase II in the transcription of mRNA-encoding genes. The emphasis will be on the mechanisms by which these factors help polymerase to recognize promoters, open up the DNA around the transcriptional start site, and transition from the initiation to the elongation phase. The general factors that assist RNA polymerase II will be compared to those that assist the other two eukaryotic RNA polymerases (RNA polymerases I and III) and to those that assist the archaeal RNA polymerase.

Chapter 4: Mechanisms of transcriptional activation

This chapter examines the mechanisms by which activators stimulate the basal machinery and emphasizes unifying themes across the domains of life. Through the presentation of examples from both bacteria and eukaryotes, it will be shown that direct contacts between regulatory factors and the basal machinery stimulate transcriptional initiation. The selected examples illustrate the ability of different activators to accelerate different steps in the transcription cycle, from the binding of RNA polymerase to the promoter, to the opening up of the promoter, to the transition of polymerase from the initiation to the elongation phase.

Chapter 5: Transcriptional control through the modification of chromatin structure

Eukaryotes have developed a mode of transcriptional regulation not available in bacteria that employs a highly developed set of coactivators and corepressors to modify chromatin structure. Chapter 5 introduces the coregulators that direct covalent and non-covalent changes in chromatin structure and examines the mechanistic connections between histone modification state and transcriptional activity.

Chapter 6: Epigenetic control of transcription

This chapter introduces the concept of the epigenetic state. Several examples of epigenetic regulation are presented to illustrate unifying mechanisms such as the use of non-coding RNA and positive feedback loops to establish and maintain epigenetic states. These examples demonstrate the broad conservation of epigenetic mechanisms across the eukaryotic domain, reflecting the central role of epigenetics in the developmental regulation of transcription.

Chapter 7: Combinatorial control in development and signal transduction

This chapter introduces combinatorial control, by which multiple regulatory factors interact to determine transcriptional states. Examples are selected to illustrate how the two basic forms of combinatorial control, namely synergy and antagonism, can be used to refine patterns of transcription during development and to elicit sophisticated responses to environmental cues.

PROBLEMS

1 What are some possible advantages of using one set of factors (the basal machinery) to recognize promoters and catalyze transcription, and a separate set of factors (the regulatory machinery) to control transcription?

2 Can you think of mechanisms by which CRMs might act over long distances other than DNA looping?

FURTHER READING

RNA biogenesis

Bentley, D.L. (2005) Rules of engagement: co-transcriptional recruitment of pre-mRNA processing factors. *Curr Opin Cell Biol*, **17**, 251–256. *A review article exploring the idea that RNA polymerase II is an mRNA factory.*

Distant enhancers and DNA looping

Banerji, J., Rusconi, S. and Schaffner, W. (1981) Expression of a beta-globin gene is enhanced by remote SV40 DNA sequences. *Cell*, **27**, 299–308. *The discovery of eukaryotic enhancers.*

Ptashne, M. (1988) How eukaryotic transcriptional activators work. *Nature*, **335**, 683–689. *A review article discussing some of the various ways in which enhancers might signal to distant promoters.*

Tolhuis, B., Palstra, R.J., Splinter, E., Grosveld, F. and de Laat, W. (2002) Looping and interaction between hypersensitive sites in the active beta-globin locus. *Mol Cell*, **10**, 1453–1465. *Demonstration that promoters and distant enhancers can be brought together inside the nucleus by DNA looping.*

The domains of life and evolution of the transcriptional machinery

Woese, C.R. and Fox, G.E. (1977) Phylogenetic structure of the prokaryotic domain: the primary kingdoms. *Proc Natl Acad Sci USA*, **74**, 5088–5090. *The discovery, based on the analysis of 16S rRNA sequences, that there are two completely distinct groups of prokaryotic organisms.*

Iwabe, N., Kuma, K., Hasegawa, M., Osawa, S. and Miyata, T. (1989) Evolutionary relationship of archaebacteria, eubacteria, and eukaryotes inferred from phylogenetic trees of duplicated genes. *Proc Natl Acad Sci USA*, **86**, 9355–9359. *Molecular phylogenetic analysis showing that the archaea and eukaryotes are more closely related to one another than they are to bacteria.*

Levine, M. and Tjian, R. (2003) Transcription regulation and animal diversity. *Nature*, **424**, 147–151. *A review article that makes the argument that evolutionary diversity is primarily due to divergence of transcriptional regulatory networks rather than to divergence of protein-coding sequences.*

2

RNA polymerases and the transcription cycle

Key concepts

- The transcription cycle consists of initiation, elongation, and termination phases
- The structure of RNA polymerase reveals how this molecular machine coordinates the phases of transcription

2.1 INTRODUCTION

At its most basic level, transcription is simply the synthesis of a strand of RNA using one of the strands of a DNA duplex as a template. The enzymatic machine responsible for this process is RNA polymerase, which moves down a DNA duplex transiently denaturing the DNA as it goes, using one of the strands in the denatured bubble as the template for the synthesis of RNA. RNA strands are always elongated in the 5′ to 3′ direction, and, as a consequence of the antiparallel nature of nucleic acid duplexes, the template strand in the DNA is always read in the 3′ to 5′ direction.

The transcription cycle can be divided into an initiation phase, an elongation phase, and a termination phase (Figure 2.1), and an understanding of the features of each of these phases is essential to an understanding of transcriptional regulation. During initiation, the RNA polymerase recognizes and binds special sequences near the start of genes termed promoters and begins the synthesis of new RNA strands. In the elongation phase, RNA polymerase moves down the template, adding nucleotide residues to the 3′ end of the RNA strand in a DNA template-directed manner. The complex between RNA polymerase, the DNA template, and the nascent RNA strand is termed the ternary elongation complex (TEC). An important property of the TEC is its extreme processivity – once TEC formation occurs, RNA polymerase rarely releases the template or the transcript until a termination signal is encountered. Termination then ensures the timely release of the transcript and the template allowing the RNA polymerase to be reutilized for another cycle of transcription.

The major phases and subphases of the transcription cycle as well as the RNA polymerases are conserved across all three domains of life. Therefore, the discussion in this chapter will integrate information about bacteria, archaea, and eukaryotes. The focus will frequently be on bacteria because the bacterial transcription cycle has been studied longest and is understood in greatest detail. The bacterial transcription cycle and polymerases will, however, be frequently compared to their counterparts in archaea and eukaryotes to highlight similarities and differences.

2.2 CORE RNA POLYMERASES

RNA polymerases are of two basic varieties, the multisubunit polymerases and the single subunit polymerases. The multisubunit enzymes are encoded by and responsible for the transcription of the bacterial, archaeal, and eukaryotic genomes, while the single subunit enzymes are encoded by and responsible for the transcription of some bacteriophage genomes (e.g., T7 bacteriophage). Although the single subunit and multisubunit enzymes exhibit similar catalytic properties, they evolved

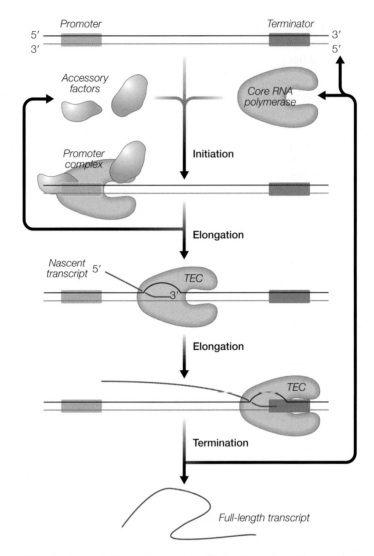

Figure 2.1 *The transcription cycle consists of initiation, elongation, and termination phases.* See text for explanation.

independently of one another. This chapter will focus exclusively on the multisubunit RNA polymerases.

In discussing the multisubunit polymerases, the term "core polymerase" is used to denote the set of subunits required for accurate template-directed elongation of an RNA strand. These core polymerases are not sufficient for initiation, which usually requires additional subunits or accessory factors. Both elongation and termination are also sometimes stimulated by accessory factors.

2.2.1 Bacterial core polymerases

In essentially all bacteria, the core polymerase contains three distinct gene products termed, in order of increasing size, α, β, and β'. The catalytically active unit contains two copies of the α chain and thus the subunit structure is $\alpha_2\beta\beta'$. In addition, core polymerases often contain an additional subunit termed ω. This small polypeptide helps to stabilize the enzyme, but is not essential for catalysis (and indeed the gene that encodes ω in *Escherichia coli* is not required for viability). The X-ray crystal structure of the core polymerase from thermophilic bacteria shows that the two α subunits form a symmetric homodimer and anchor the β and β' subunits together, giving the protein the appearance of a crab claw (Plate 2.1A, B).

> **Thermophilic bacteria** – bacteria, such as *Thermus aquaticus*, adapted to grow at high temperatures. The proteins from these organisms are of high stability to permit function at high temperature. As a result, they often yield high quality crystals and are therefore favored for structure determination by X-ray crystallographers.

2.2.2 Eukaryotic and archaeal core polymerases

While bacteria contain a single core RNA polymerase used in essentially all transcription, eukaryotes contain three distinct core RNA polymerases, named RNA polymerase I (Pol I), RNA polymerase II (Pol II), and RNA polymerase III (Pol III). These three enzymes exhibit differential sensitivity to an inhibitor termed α-amanitin, a naturally occurring cyclic octapeptide (Figure 2.2A). The fungus that produces this toxin, *Amanita phalloides* (the "deathcap mushroom"), is the most toxic of all poisonous mushrooms thus testifying to the fundamental importance of RNA polymerase. Different polymerases exhibit different sensitivity to this compound: mammalian Pol II is inhibited by nanomolar concentrations of α-amanitin, mammalian Pol III is inhibited by micromolar concentrations of α-amanitin, and mammalian Pol I is essentially resistant to α-amanitin. By determining the sensitivity of various transcription units to α-amanitin, it was found that Pol II transcribes mRNA-encoding genes, Pol III transcribes the genes encoding tRNA and 5s rRNA, and Pol I transcribes the gene encoding the

> **Why are archaeal and eukaryotic core RNA polymerases so much more complex than their bacterial counterparts?**
>
> Given the relative complexity of eukaryotic genomes, it is tempting to argue that eukaryotic RNA polymerase complexity reflects a need for complex patterns of gene regulation. However, archaeal polymerase is as complex as eukaryotic Pol II despite the relative simplicity of archaeal genomes. Thus, there is no simple correlation between genome complexity and core RNA polymerase complexity.

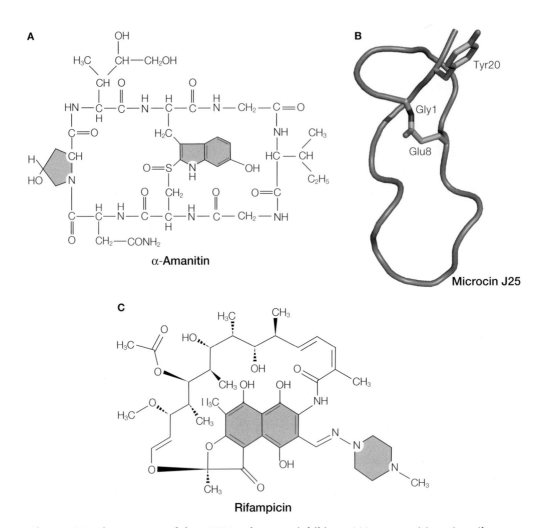

Figure 2.2 *The structures of three RNA polymerase inhibitors.* (A) α-amanitin primarily inhibits RNA polymerase II, and, to a lesser extent, RNA polymerase III. It is a cyclic octapeptide containing hydroxylated and normal amino acids, as well as an unusual covalent linkage between a hydroxyltryptophan side chain and an oxidized cysteine side chain. (B) Microcin J25 is an inhibitor of *E. coli* RNA polymerase. It is a 21 residue long peptide in which an amino linkage between the nitrogen of Gly1 and the side chain carboxyl group of Glu8 results in the formation of a lariat. This is a backbone trace in which only the side chains of Glu8 and Tyr20 are shown. The C-terminal end of the polypeptide passes through the protein loop and is prevented from slipping out of the loop by the bulky Tyr20 side chain. Because the C-terminal end passes through the loop in this way, the structure is sometimes described as a "lariat protoknot" (PDB id 1PP5: Bayro, M.J., Mukhopadhyay, J., Swapna, G.V.T., Huang, J.Y., Ma, L.-C., Sineva, E., Dawson, P.E., Montelione, G.T. and Ebright, R.H. (2003) Structure of antibacterial peptide microcin J25: a 21-residue lariat protoknot. *J Am Chem Soc,* **125**, 12382–12383.). (C) Rifampicin is an inhibitor of many bacterial polymerases.

large rRNA precursor (45s pre-rRNA), which is ultimately cleaved into three distinct rRNA species (5.8s, 18s, and 28s rRNA). Genes transcribed by Pol I, Pol II, and Pol III are termed class I, class II, and class III genes, respectively.

The eukaryotic core RNA polymerases exhibit greater complexity than their bacterial counterparts, containing 12–16 distinct polypeptide chains each (Table 2.1), and the same is true for the single archaeal core polymerase. The bacterial, eukaryotic, and archaeal enzymes have clearly descended from a common ancestor since sequence analysis reveals that all the subunits of the bacterial core polymerase have counterparts in the eukaryotic and archaeal enzymes (Table 2.1). The largest subunit of each enzyme is homologous to the largest subunit of the bacterial enzyme (β'), while the second largest subunit of each enzyme is homologous to the second largest subunit of the bacterial enzyme (β). The α_2 homodimer found in the bacterial enzyme is replaced in each eukaryotic and archaeal enzyme by a heterodimer of two subunits with sequence similarity to α. In accord with this sequence homology, the X-ray crystal structure of yeast Pol II reveals significant structural similarity to the bacterial enzyme (Plate 2.2).

The functions of the eukaryotic and archaeal RNA polymerase subunits for which no counterparts are found in bacteria are not well understood. In some cases, they may serve architectural roles, helping to stabilize the enzyme, while in other cases they may serve regulatory roles, mediating interactions with the regulatory machinery.

The observation that archaeal core RNA polymerase is more closely related to the eukaryotic than to the bacterial polymerase provides support for the belief that archaea and eukaryotes share a common ancestor not shared by bacteria. Based on its subunit composition, the archaeal polymerase seems to be more closely related to Pol II than to Pol I or Pol III. Pol II may therefore be ancestral to Pol I and Pol III.

The Pol II CTD

After the evolutionary split between archaea and eukaryotes, RNA polymerase II acquired a feature that is not found in any other RNA polymerase, namely the C-terminal domain or CTD. This is a domain found at the C-terminal end of the largest subunit (the β'-homologous subunit) of Pol II (Figure 2.3A). It consists of multiple tandem imperfect repeats of a seven amino acid sequence (consensus = YSPTSPS), with the number of repeats varying from species to species (e.g., there are 26 repeats in yeast, 45 in *Drosophila*, and 52 in humans). The CTD is invisible in the X-ray crystal structures of Pol II due to the relatively unstructured nature of this Pol II "tail". Two of the seven residues in the seven amino acid sequence (the serines at positions 2 and 5) frequently serve as acceptor sites for phosphate groups, and cells contain a wide array of protein kinases that

Table 2.1

Subunit structure of the multisubunit RNA polymerases[1]

Pol II[2] (Saccharomyces cerevisiae)	Pol I[2] (Saccharomyces cerevisiae)	Pol III[2] (Saccharomyces cerevisiae)	Bacteria (Escherichia coli)	Archaea (Methanococcus jannashii)
Rpb1 (1733)	Rpa190 (1664)	Rpc160 (1460)	β′ (1407)	A′ (1341) + A″ (859)[3]
Rpb2 (1224)	Rpa135 (1203)	Rpc128 (1149)	β (1342)	B′ (636) + B″ (498)[3]
Rpb3 (318)	Rpc40 (335)	Rpc40 (335)	α (329)	D (191)
Rpb4 (221)[4]				F (102)[4]
Rpb5 (215)	Rpb5 (215)	Rpb5 (215)		H (78)
Rpb6 (155)	Rpb6 (155)	Rpb6 (155)	ω (91)	K (57)
Rpb7 (171)[4]	Rpa43 (326)[4]	Rpc25 (212)[4]		E (187)[4]
Rpb8 (146)	Rpb8 (146)	Rpb8 (146)		
Rpb9 (122)	Rpa12 (125)	Rpc11 (110)		TFS (108)[5]
Rbp10 (70)	Rbp10 (70)	Rbp10 (70)		N (73)
Rbp11 (120)	Rpc19 (142)	Rpc19 (142)	α (329)	L (99)
Rbp12 (70)	Rbp12 (70)	Rbp12 (70)		P (46)
	Rpa49 (415)			
	Rpa34 (233)			
	Rpa14 (137)			
		Rpc82 (654)		
		Rpc53 (422)		
		Rpc37 (282)		
		Rpc34 (317)		
		Rpc31 (251)		
		Rpc17 (161)[4]		

[1]Shading indicates subunits that are shared between two or more polymerases. Subunits in the same row have significant sequence homology. Numbers in parentheses indicate the lengths of the subunits in amino acid residues.

[2]Eukaryotic RNA polymerase subunit nomenclature: subunits of RNA polymerase II are named Rpb1 through Rpb12 in approximate order of their size. The eukaryotic subunits not found in RNA polymerase II have names beginning with Rpc for subunits found in Pol III or in both Pol I and Pol III, or beginning with Rpa for subunits found only in Pol I. The number following Rpa or Rpc refers to the apparent molecular mass of the subunit in kilodaltons as determined by SDS-PAGE.

[3]The two largest subunits in the eukaryotic and bacterial polymerases are each split into two polypeptides in the archaeal polymerase.

[4]Rpb4 and Rpb7 form a heterodimer that constitutes a dissociable, non-essential module of Pol II. This is clearly homologous to the E/F module found in the archaeal polymerase. While Pol I and Pol III both lack factors with sequence homology to Rpb4, the Pol III subunit Rpc17 heterodimerizes with Rpc25, the Pol III subunit with homology to Rpb7. Rpc17 may therefore be the functional equivalent of Rpb4.

[5]TFS is identified here as an archaeal RNA polymerase subunit based on its significant similarity to Rpb9, Rpa12, and Rpc11 and its ability to bind the archaeal polymerase. TFS also exhibits homology to the eukaryotic elongation factor TFIIS, a factor that increases elongation efficiency by Pol II, but which is not stably associated with Pol II. In terms of its functionality, TFS may be more like TFIIS than Rpb9: it binds TFIIS and stimulates elongation, but it is not required for elongation. Therefore, its status as an RNA polymerase subunit is debatable.

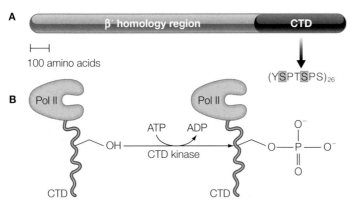

Figure 2.3 *The primary structure and phosphorylation of the yeast Pol II large subunit (Rpb1).* (A) The first ~1500 amino acids of this subunit exhibit extensive similarity to the β′ subunit of the bacterial enzyme, while the remaining ~200 amino acids consists primarily of imperfect repeats of the indicated seven amino acid sequence. In yeast, this seven amino acid sequence is repeated 26 times. Ser2 and Ser5 of this sequence are boxed to indicate that they frequently serve as phosphoacceptor sites for C-terminal domain (CTD) kinases. (B) Ser2 and Ser5 are subject to phosphorylation by a number of CTD kinases.

catalyze this phosphorylation reaction (Figure 2.3B). As we will see in the next chapter, the phosphorylation of the Pol II CTD is intimately connected to the transcription cycle and to the overall process of RNA biogenesis.

2.3 TRANSCRIPTIONAL ELONGATION

2.3.1 Phosphoester linkage formation

The basic reaction

The basic reaction catalyzed by RNA polymerase is the addition of a nucleotide residue to the 3′ end of a nascent RNA strand. The overall reaction for a single step in the elongation cycle is:

$$(RNA)_n + NTP \rightarrow (RNA)_{n+1} + \text{pyrophosphate,}$$

where $(RNA)_n$ is an n nucleotide-long piece of RNA and NTP (nucleoside triphosphate) can be ATP, GTP, CTP, or UTP.

   In this reaction, the phosphoanhydride linkage between the α- and β-phosphate groups in an NTP is broken and replaced with a phosphoester linkage to the newly added nucleotide residue (Figure 2.4). The reaction is energetically favorable, having an equilibrium constant of about 100.

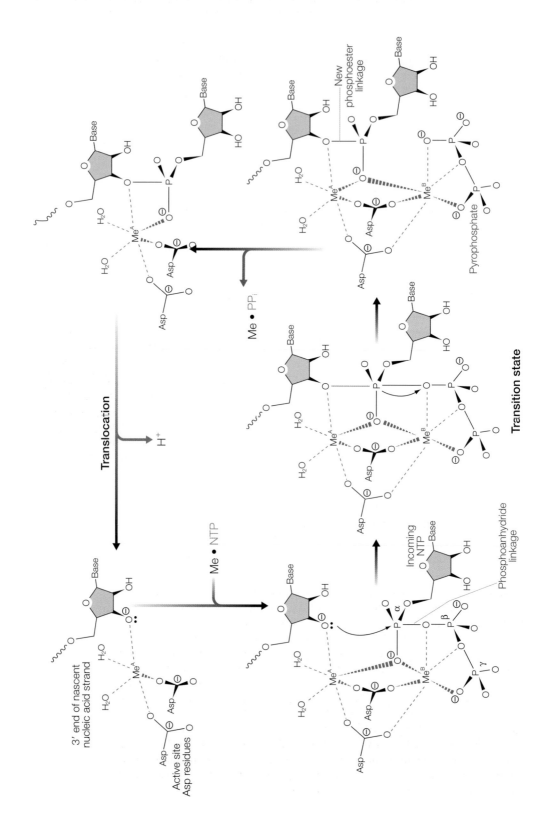

This means that the reaction will proceed in the forward direction as long as the ratio of pyrophosphate to NTP is less than about 100. If the ratio exceeds this value, the RNA polymerase will begin to operate in reverse to degrade rather than synthesize RNA. In cells, pyrophosphate and nucleotide concentrations are roughly equal (~1 mM each) and so the enzyme operates in the synthetic rather than the degradative mode.

In each round of polymerization, the 3′ hydroxyl group at the end of the growing RNA strand attacks the α-phosphorus atom on the NTP (Figure 2.4). This leads to a transition state in which the α-phosphorus atom is covalently bonded to five oxygen atoms. The transition state decays to the product by the cleavage of the phosphoanhydride linkage between the α- and β-phosphate groups to release the pyrophosphate leaving group.

Catalysis by metal ions

All nucleic acid polymerases (both DNA and RNA polymerases) rely on common mechanisms to achieve catalysis. All such enzymes are thought to possess two metal ions (usually Mg^{2+}) in the active site that are held in position by coordinate bonds to aspartate side chains in the enzyme (Figure 2.4, Box 2.1). These Mg^{2+} ions also make coordinate bonds with oxygen atoms in the substrates. The distances and angles between the Mg^{2+} and the oxygen atoms in the substrates are only optimal for coordinate bond formation when the reaction is in the transition state. These bonds to the Mg^{2+} ions therefore lower the free energy of the transition state relative to that of the substrates. This lowers the activation energy and accelerates the reaction (Box 2.2).

> **Coordinate bonds** – a type of covalent bond that often forms between a metal ion and an electronegative atom such as oxygen or nitrogen. They are weaker than conventional covalent bonds, but stronger than ionic bonds.

Figure 2.4 (*opposite page*)*Catalysis of the elongation cycle in RNA biosynthesis by metal ions.* The substrates and products are shown in black, while the enzyme side chains, catalytic metal ions, and tightly bound water molecules are shown in purple. Multiple aspartate side chains form coordinate bonds (dashed blue lines) to the metal ions positioning them for catalysis. The contact between metal A and the 3′ hydroxyl group lowers its pKa, easing the ionization of this group and therefore making it a more potent nucleophile. In the transition state, the α-phosphorus atom of the nucleoside triphosphate (NTP) is transiently bonded to five oxygen atoms. The geometry of the coordinate bonds is optimal in the transition state, thereby stabilizing the transition state relative to the substrates and lowering the activation barrier to the reaction. Note that metal B makes numerous coordinate bonds to the phosphate groups in the incoming NTP and is thought to enter the active site along with the NTP.

Box 2.1

Evidence for metal ions in the RNA polymerase active site

How do we know that metal ions are required for catalysis by RNA polymerases? This idea is based in part on X-ray diffraction experiments demonstrating the presence of metal ions in the active site. This is not definitive by itself, since it is possible that the metal ions only bound during the crystallization process. Additional support comes from studies of *E. coli* RNA polymerase in solution in which iron ions (Fe^{2+}) were allowed to displace the Mg^{2+} bound to the polymerase. Unlike magnesium, iron is a redox active metal meaning it can readily shed electrons, which can then react with hydrogen peroxide to produce highly reactive hydroxide radicals (Figure B2.1A). These radicals, in turn, react very quickly with

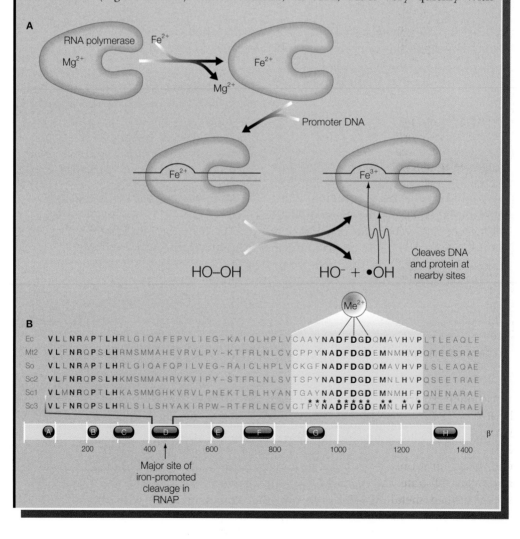

nearby macromolecules, sometimes generating breaks in the protein or nucleic acid backbone. Because the hydroxide ion is so unstable, cleavage generally occurs only at sites on the macromolecules within a few angstroms of the iron ion. The Fe^{2+} in the iron-substituted RNA polymerase was found to promote the cleavage of the β′-strand of the polymerase at a site very close to three aspartate residues that are highly conserved in all multisubunit RNA polymerases, including the three eukaryotic polymerases, bacterial polymerases, and archaeal polymerases (Figure B2.1B). In addition, when the iron-substituted enzyme was bound to a promoter, the iron was found to cleave the promoter right at the site where transcription would normally initiate. These findings strongly suggest that metal ions are held in the active site by conserved aspartate residues.

Figure B2.1 (*opposite page*)*Mapping metal-binding sites in* E. coli *RNA polymerase.* (A) To map metal-binding sites in the RNA polymerase and bound template, the enzyme was incubated with excess Fe^{2+} allowing iron ions to displace magnesium ions in the polymerase. The iron-bound enzyme was then bound to a promoter. Reduction of hydrogen peroxide (HO–OH) by the bound iron results in the formation of hydroxide radicals near the metal-binding site. This highly labile species reacts almost immediately with whatever is on hand, sometimes cleaving the nearby polypeptide or nucleic acid backbone. Preferred positions of cleavage in the protein and DNA are then determined by measuring the sizes of the cleavage products. (B) Cleavage of RNA polymerase by the bound iron occurs within a region of the β′ subunit containing three highly conserved aspartate residues. The bar at the bottom is a schematic representation of the primary structure of the β′ subunit. Eight regions (labeled A–H) that are highly conserved from bacteria to archaea to eukaryotes are indicated. The cleavage site maps to region D. The sequences of region D from *E. coli*, *Methanobacterium thermoautotrophicum* (an archaea), *Spinacia oleracea* chloroplast, and *Saccharomyces cerevisiae* Pol I, Pol II, and Pol III are aligned above the bar. A highly conserved region containing an absolutely conserved NADFDGD motif is boxed. Mutations within this motif inactivate the polymerase. It was thus concluded that the asparates in the conserved motif are responsible for coordinating catalytic metal ions. Later determination of the RNA polymerase crystal structure confirmed this hypothesis. B, from Zaychikov, E. et al. (1996) *Science*, **273**, 107–109, with permission from AAAS.

Although a variety of studies strongly support the idea that catalysis by RNA and DNA polymerases requires two Mg^{2+} ions in the active site, many X-ray crystal structures of RNA polymerases (e.g., Plates 2.1 and 2.2) show only a single metal ion in the active site. This is because the second metal ion is only stably bound to the enzyme in the presence of the nucleotide substrate. The metal ion that is permanently bound in the active site is termed metal A, while the one that comes in along with the nucleotide is termed metal B (Figure 2.4).

Box 2.2

Transition state theory

An understanding of how enzymes, including RNA polymerases, are able to speed up a reaction requires a basic understanding of transition state theory. In any chemical reaction, the pathway from reactants (or substrates in the case of enzyme catalyzed reactions) to products proceeds through an unstable high free energy state termed the transition state. This is often illustrated with the use of a transition state diagram (Figure B2.2) in which the *x*-axis represents the progress of the reaction, while the *y*-axis represents the free energy. The so-called activation free energy (or "activation barrier") of the reaction is simply the difference between the free energy of the transition state and the free energy of the reactants. This activation free energy ($\Delta G^{\ddagger}$) is related to the rate constant of the reaction (k) by the following equation:

$$k = Ae^{-\Delta G\ddagger/RT}$$

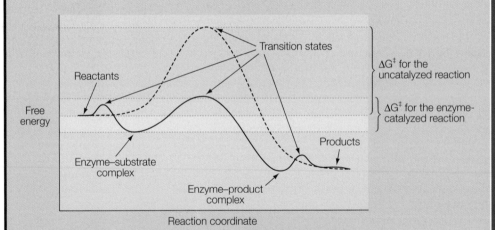

Figure B2.2 *A transition state diagram.* This diagram compares a hypothetical uncatalyzed reaction (dashed line) to its enzyme-catalyzed counterpart (solid line). The uncatalyzed reaction occurs in a single step proceeding through a single transition state. In contrast, the enzyme-catalyzed reaction occurs in multiple steps proceeding through multiple transition states. These steps include binding of the substrates to the enzyme, conversion of the substrates to the products, and release of the products from the enzyme. The activation free energy $\Delta G^{\ddagger}$ for conversion of substrates to products is less for the enzyme-catalyzed reaction than for the uncatalyzed reaction because the enzyme stabilizes the transition state relative to the substrates.

where A is a proportionality constant, R is the gas constant, and T is the temperature. The negative sign in front of $\Delta G^{\ddagger}$ means that the larger the activation free energy, the slower the reaction. Catalysts such as enzymes accelerate reactions by lowering the free energy of the transition state relative to the free energy of the reactants thereby reducing the activation barrier.

To apply these concepts to RNA polymerase, we simply take into account the fact that the geometry of the coordinate bonds between the metal ions and the reactants is only optimal in the transition state. Thus, these coordinate bonds lower the free energy of the transition state significantly more than they lower the free energy of the unreacted substrates thus reducing the activation barrier.

The transcription bubble

The DNA template precisely directs the selection of NTPs during elongation. In the TEC, the template DNA contains a denatured bubble to which the final eight or nine nucleotides of the nascent transcript are basepaired, forming a heteroduplex. The denatured bubble extends a few nucleotides beyond the heteroduplex at either end and hence the bubble is about 13 bp in length (Figure 2.5, Plate 2.3D). At the beginning of each step of elongation, the 3′ end of the nascent transcript is positioned in

> **Heteroduplex** – a double-stranded nucleic acid in which one strand is DNA and the other strand is RNA.

the catalytic site. A nucleotide then binds to the enzyme making a Watson–Crick basepairing interaction with the next exposed nucleotide in the transcription bubble. Metal ions in the active site then catalyze the formation of a new phosphoester linkage. Before the next round of nucleotide addition can occur, the polymerase translocates down the template. As a consequence, the denatured bubble moves 1 bp down the template and the new 3′ end is repositioned in the active site to allow the process to repeat itself (Figure 2.5).

2.3.2 Features of the ternary elongation complex

A model for the structure of the bacterial TEC based on protein–nucleic acid cross-linking studies (Plate 2.3) suggests functions for a number of the structural features visible in the X-ray crystal structure of the core polymerase (Plates 2.1, 2.3). These features include the primary channel, the secondary channel, the β-flap, and the rudder. The remainder of this section will discuss the functions of these four features in the elongation process. The discussion of these RNA polymerase features is primarily based upon studies carried out on the bacterial elongation

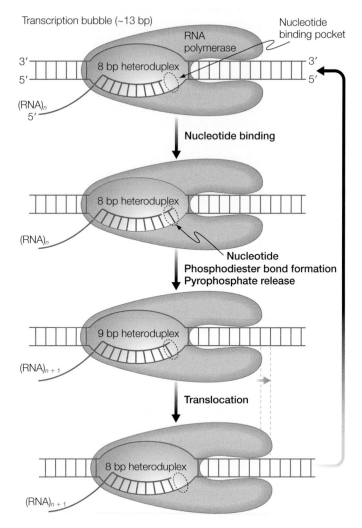

Figure 2.5 *The transcription bubble and heteroduplex in RNA polymerase.*
See text for explanation.

complex. However, these features are conserved in eukaryotic RNA polymerases and therefore the conclusions presented here also apply to the eukaryotic enzymes.

The primary channel and the approach of DNA to the active site

As we have seen, RNA polymerase has the overall shape of a crab claw – the upper claw consists primarily of the β subunit, while the lower claw consists primarily of the β′ subunit (Plates 2.1, 2.2, 2.3). Highly conserved aspartate residues at the back of the 27 Å wide channel (the primary channel) between the two claws coordinate a magnesium ion which is thought to represent metal A. The TEC model indicates that the DNA template runs through this channel (Plate 2.3). To reach the magnesium ion, the DNA must be severely bent – the flexibility needed to allow for this bending is provided by the single-stranded DNA in the transcription bubble (single-stranded DNA is much less rigid than double-stranded DNA) (Plate 2.3D).

The X-ray crystal structure of the core polymerase in complex with the antibiotic rifampicin (Figure 2.2C) strongly suggests that the proposed position of the transcription bubble at the back of the primary channel near the magnesium ion is correct. Rifampicin, a potent inhibitor of bacterial RNA polymerases, is produced by the bacterium *Nocardia mediterranei*, and is an important therapeutic agent in the treatment of tuberculosis. It binds to the core polymerase about 12 Å from metal A. In this position, the TEC model predicts that the antibiotic will physically block elongation once the transcript is more than a few nucleotides in length, a prediction that is in good accord with experimental findings.

The secondary channel: entry site for the nucleotides

During elongation, the only way for nucleotides to reach the active site is by diffusing in through a pore in the core polymerase known as the secondary channel (Plate 2.3C). This is because the DNA blocks access to the active site through the primary channel. An antibiotic termed microcin J25 is thought to inhibit RNA polymerase by binding to the secondary channel and blocking access to the active site. This antibiotic, which has the form of a 21 amino acid residue-long peptide lariat (Figure 2.2B), is produced by some strains of *E. coli* and is a potent inhibitor of *E. coli* RNA polymerase. Strains that produce microcin J25 also produce an efficient microcin J25 exporter, allowing them to deliver the antibiotic to their competitors. By killing competing strains of *E. coli*, this antibiotic gives the strain that produces it a selective advantage. As expected, mutations in RNA polymerase that render the enzyme resistant to this antibiotic all map to residues lining the secondary channel.

The β-flap and the RNA exit channel

As the 5′ end of the transcript exits the transcription bubble, its path takes it underneath the β-flap (Plates 2.1C and 2.3B). This is a structured protein loop of the β subunit that is only loosely anchored to the rest of the polymerase. The RNA exit channel beneath the flap accommodates about 10 nucleotide residues. Thus, since the heteroduplex in the active site is about 8–9 bp in length, the 5′ end of the transcript only encounters the surrounding medium once the transcript is about 20 nucleotides in length. As will be discussed in later sections, the presence of the RNA in this exit channel has important implications for both initiation and termination.

The rudder: a role in processivity

> **Processivity** – a term used to describe an enzyme that catalyzes multiple cycles of catalysis without releasing the intermediates between the rounds of catalysis. A processive polymerase is one that adds multiple residues to the end of a polymer without releasing the growing polymer between the cycles of addition.

RNA polymerases are highly processive enzymes. After extending an RNA strand by one nucleotide, they do not generally release the RNA strand, but rather they move along the RNA and associated DNA template positioning the new 3′ end in the active site in preparation for another round of catalysis. Indeed, these enzymes rarely release the transcript before encountering a termination signal. This extreme processivity indicates that the polymerase binds the transcript with high affinity throughout the elongation phase. This high affinity results, in part, from basepairing interactions between the transcript and the template in the transcription bubble. However, this heteroduplex is not sufficient to account for the stability of the elongation complex – additional stabilizing interactions between the transcript and the protein are required. These stabilizing contacts are provided in part by the rudder. This is a protein loop that projects from the floor of the active site (Plate 2.3B). It is thought to bind the transcript as it exits the bubble, thereby stabilizing the association of the RNA with the TEC.

2.3.3 RNA polymerase as a motor

> **What is the molecular basis of translocation?**
>
> In other words, how does RNA polymerase move along the template in one direction? The equilibrium sliding model suggests that translocation may simply be the result of Brownian motion coupled to energetically favorable phosphoester linkage formation. However, it is also possible that the elongation cycle is coupled to conformational changes that allow the enzyme to move down the template in an inchworm-like manner.

In moving down the DNA, RNA polymerase behaves similarly to a molecular motor such as myosin (a protein that moves along actin filaments to produce force during muscle contraction). However, unlike myosin, which exists solely to generate motion and force, RNA polymerase exists to synthesize RNA – its ability to function as a molecular motor is a byproduct of the need for processive RNA synthesis. Surprisingly, however, RNA polymerase is at least as forceful a motor as myosin. This conclusion derives from experiments employing a laser beam in a device known as an optical trap to grab the end of a DNA template undergoing transcription by RNA polymerase anchored to a surface. By using the laser beam to pull the template away from the anchored polymerase, it was found that *E. coli* RNA polymerase could continue to transcribe against a force of up to 14 piconewtons, which is greater than the force generated by myosin as it moves along actin filaments.

The ability of RNA polymerase to move along the template in such a forceful manner must somehow relate to the ability of the enzyme to hold the template tightly while at the same time coupling the energetically favorable chemistry (phosphoester linkage formation) to directional movement along the template. One simple model for how this might happen is sometimes termed the equilibrium sliding model (Figure 2.6). According to this model, polymerase slides back and forth along the transcription bubble due to Brownian motion, so that template nucleotide n (the nucleotide basepaired to the residue at the 3′ end of the growing RNA strand) and template nucleotide $n+1$ (the next nucleotide to be copied) alternately occupy the catalytic site. Energetically favorable phosphoester bond formation, which can only occur when template nucleotide $n+1$ occupies the active site, then serves to drive the enzyme down the template in a directional manner.

2.3.4 Elongation factors and backtracking

Although core polymerases are, by definition, sufficient for elongation, the rate and processivity of elongation in both bacteria and eukaryotes is modulated by numerous elongation factors. One class of these factors, which includes the Gre factors in bacteria and the functionally homologous TFIIS in eukaryotes, allow polymerases to negotiate roadblocks in the template such as DNA damage or DNA-bound proteins.

When an RNA polymerase encounters such a roadblock, elongation slows down. The slowed polymerase frequently begins to move backward along the template in a process known as backtracking (Figure 2.7A). Backtracking is not the reversal of polymerization since the RNA is not degraded during backtracking. Instead, as the polymerase and transcription bubble move back toward the promoter, the 3′ end of the nascent transcript disengages from the template. Analysis of the TEC model strongly suggests that the 3′ end of the transcript in the backtracked complex must exit the active site through the secondary channel. In the backtracked configuration, an internal phosphoester bond, rather than a 3′-OH group, is positioned in the active site next to metal A. Resumption of elongation requires the hydrolytic cleavage of this phosphoester bond to release a short oligonucleotide and to generate a new 3′ end in the active site. The polymerase can then try once again to proceed through the roadblock.

Bacterial core RNA polymerase has the intrinsic ability to catalyze this hydrolysis reaction. However, a level of hydrolysis sufficient to efficiently

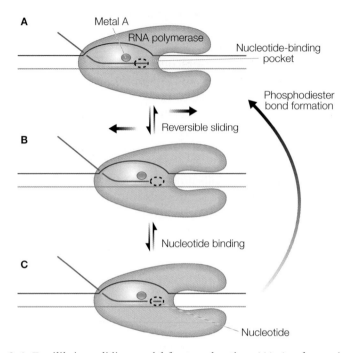

Figure 2.6 *Equilibrium sliding model for translocation.* (A) A schematic view of the elongating polymerase after a round of phosphodiester bond formation, but before translocation. (B) A view of the elongating polymerase after translocation, which opens up the nucleotide-binding site for a new incoming nucleotide. (C) Post-translocation polymerase with bound nucleotide. Brownian motion results in the oscillation of the polymerase between configurations A and B. Only in configuration B, can nucleotide binding and thus phosphodiester bond formation occur. The favorable formation of phosphodiester bonds is therefore able to drive the polymerase in the forward direction.

resolve backtracked complexes requires one of two closely related elongation factors termed GreA and GreB. It is thought that these factors have the ability to reach in through the secondary channel of the backtracked polymerase and position two highly conserved negatively charged amino acid residues near the active site. These conserved residues serve to coordinate a Mg^{2+} ion, placing it in the position normally occupied by metal B during the elongation cycle. Recall that metal B normally only enters the active site with an incoming nucleotide. In the presence of the Gre factor, however, metal B can bind in the absence of nucleotide and then metals A and B can catalyze an attack by water (or hydroxide ion) upon the phosphodiester bond in the active site, hydrolyzing the linkage and generating the new 3′ end needed to resume elongation (Figure 2.7B).

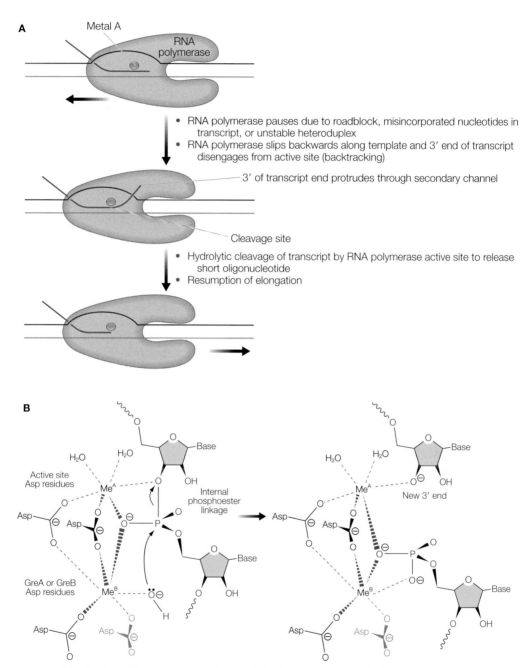

A

Metal A

RNA polymerase

- RNA polymerase pauses due to roadblock, misincorporated nucleotides in transcript, or unstable heteroduplex
- RNA polymerase slips backwards along template and 3′ end of transcript disengages from active site (backtracking)

3′ of transcript end protrudes through secondary channel

Cleavage site

- Hydrolytic cleavage of transcript by RNA polymerase active site to release short oligonucleotide
- Resumption of elongation

B

Active site
Asp residues

GreA or GreB
Asp residues

Internal phosphoester linkage

New 3′ end

Figure 2.7 *Backtracking and the resolution of the backtracked complex.* (A) A schematic view of the backtracking process. (B) Cleavage of the backtracked complex. Two metals are required for cleavage of the backtracked RNA. Metal A is held in position by RNA polymerase aspartate residues, while stimulatory factors such as GreA or GreB in bacteria reach into the active site through the secondary channel and use conserved aspartate residues to position metal B.

In eukaryotes, Pol II uses a very similar strategy to re-engage backtracked complexes that form when the polymerase encounters a roadblock in the template. In this case, hydrolysis of the backtracked complex is stimulated by TFIIS. Although TFIIS is not evolutionarily related to the bacterial Gre factors, it seems to work by a very similar mechanism, reaching in through the secondary channel and using conserved negatively charged amino acid residues to position a metal ion in the active site.

2.4 TRANSCRIPTIONAL INITIATION

2.4.1 Distinct mechanisms for promoter recognition in bacteria and eukaryotes

While core polymerases are sufficient for the elongation phase of transcription, they cannot initiate transcription in a promoter-specific fashion. The physiological recognition of promoters and the formation of TECs require one or more polypeptides in addition to the multisubunit core polymerase. In bacteria, a single additional polypeptide termed a σ factor is required for initiation. Bacteria contain multiple σ factors, each of which renders the polymerase specific for a different set of promoters (Table 2.2). σ factors bind the core polymerase with high affinity and often copurify with core polymerase at stoichiometric levels. Thus, these factors are considered to be RNA polymerase subunits and the complex of the core polymerase with a σ factor is termed an RNA polymerase holoenzyme.

While σ factors are present in all bacteria, they are essentially absent from the two other domains of life. Instead of using σ factors, archaeal and eukaryotic core polymerases employ sets of polypeptides termed general transcription factors to recognize a promoter and establish a TEC. (An exception is provided by the RNA polymerases found in chloroplasts, which do contain σ factors – but this is an exception that proves the rule since chloroplasts are descended from cyanobacteria.) Each core polymerase uses a distinct set of general transcription factors, although there is some overlap between the sets. These sets range in complexity from the two polypeptides required for promoter recognition and TEC formation by the archaeal core polymerase to the > 25 polypeptides required by eukaryotic RNA polymerase II.

Although many general transcription factors are able to bind core archaeal and eukaryotic polymerases, these interactions are usually weak. Unlike the bacterial σ factors, general transcription factors are therefore not generally considered to be subunits of RNA polymerase. σ factors will be discussed in this chapter, while a discussion of eukaryotic general transcription factors will be deferred until Chapter 3.

Table 2.2

Sigma factors of *Escherichia coli*[1]

σ factor	−35 element	Spacer	−10 element	Class of genes
σ^{70}	TTGACA	16–18	TATAAT	Housekeeping (~95% of genes)
σ^{S}	ND		CTATACT	Stationary phase
σ^{32}	CTTGAAA	11–16	CCCATnT	Heatshock/stress
σ^{F}	(A/T)GCATA	14–15	GGn(A/G)A(T/C)A(A/C)T(A/T)	Flagella/chemotaxis
σ^{E}	GAACTT	16–17	TCT(A/G)A	Extracellular functions
σ^{fecI}	ND		ND	Extracellular functions
σ^{54}	TGGCAC[2]	5	TTGC(A/T)[2]	Nitrogen metabolism

[1]The *E. coli* genome encodes seven sigma factors – each one combines with the core polymerase to produce a holoenzyme with distinct promoter specificity. Note that the −35 and −10 sequences shown are consensus sequences. This means that they were generated by aligning multiple promoters and determining the most common nucleotide(s) at each position. Few promoters match the consensus sequences perfectly – most differ from the consensus sequences at multiple positions. In general, however, the better a promoter matches the consensus sequences, the more efficiently it is utilized by the cognate holoenzyme. The first six σ factors in this table all show significant homology to one another and define the so-called σ^{70} class of σ factors. σ^{54}, in contrast, shows little homology to the other σ factors and therefore defines a class of its own. ND, not determined.
[2]The conserved promoter elements in σ^{54} promoters are centered at about −24 and −12 rather than at −35 and −10.

2.4.2 Bacterial σ factors and promoter specificity

Initiation sequence of events

In bacterial genes, the promoter sequences required for initiation by RNA polymerase are generally located within the ~50 bp region upstream of the transcriptional start site. In particular, this region contains two moderately conserved elements, one located at about 35 bp upstream of the start site (the −35 element) and one located at about 10 bp upstream of the start site (the −10 element or Pribnow box) (Figure 2.8A). The consensus sequences within the −35 and −10 elements and the preferred spacing between these elements are determined by the particular σ factor associated with the core polymerase (Table 2.2).

The formation of the TEC from the free polymerase and DNA containing a promoter occurs in four phases (Figure 2.8B–E). First, the enzyme recognizes and binds the promoter resulting in the formation of the closed promoter complex. Second, the enzyme–promoter complex undergoes a conformational change to form the open complex. During open complex formation, the template around the transcriptional start site (from about

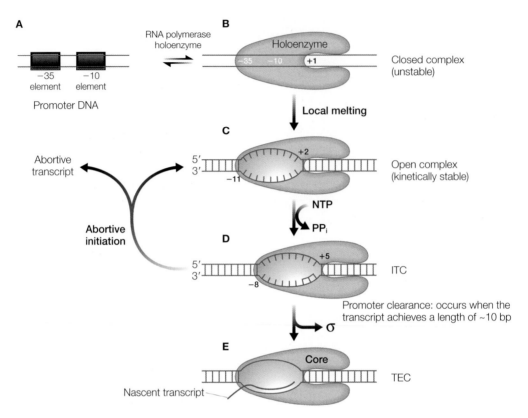

Figure 2.8 *The sequence of events leading to formation of the ternary elongation complex (TEC).* See text for explanation.

−11 to about +2) melts exposing the template strand for decoding. Third, the polymerase begins to synthesize RNA. This initial transcribing complex (ITC) lacks the processivity associated with the authentic TEC. In a process referred to as abortive initiation, short transcripts (2–10 nucleotides) are repeatedly released and after each such release the polymerase cycles back to the open complex. In the fourth and last phase of the initiation sequence, called promoter clearance, the polymerase escapes the promoter and begins to move down the template. An important feature of this step is the stabilization of the association between the nascent transcript and the enzyme resulting in the formation of the processive TEC. Both the release from the promoter and the stabilization of the complex occur when the transcript reaches about 10 nucleotides in length.

Orchestration of the bacterial initiation sequence by σ factors

The general features of the initiation sequence outlined above can be generalized to all RNA polymerases. In bacteria, this entire process is orchestrated by σ factors. Structural and functional studies from a number of labs on the bacterial holoenzyme have shed tremendous insight into how σ factors are able to guide bacterial polymerases through the initiation sequence.

σ factors contain four separate structural domains termed σ_1 through σ_4 (σ_1 is mobile and therefore invisible in the X-ray crystal structure of the holoenzyme). The structure of the bacterial holoenzyme reveals that σ binds to core polymerase through an extensive series of interactions between σ_2, σ_3, and σ_4, and the upstream face of the core polymerase (Plate 2.4A, B). Remarkably, σ_3 and σ_4 bind to the core polymerase on opposite sides of the β-flap. As discussed above, the transcript is thought to pass underneath this flap as it exits the active site during elongation. The linker between σ_3 and σ_4 also runs underneath the flap, occupying the same space that is later assumed by the nascent transcript.

Based on the structure of the holoenzyme as well as many years of functional analysis, initiation is thought to involve the following sequence of events (Figure 2.9):

1 **Closed complex formation.** The initial binding of the holoenzyme to the promoter to form the closed complex is completely mediated by the σ factor. σ_2 binds to the −10 region, while σ_4 binds to the −35 region. At this stage, the DNA lies completely outside of the primary channel (Figure 2.9A).

2 **Open complex formation.** σ_2 then initiates open complex formation by inducing the melting of the DNA in the −10 region. Although it is not clear how DNA melting is induced, one possibility is that "breathing" of the AT-rich DNA in the −10 region results in the transient opening of the DNA, and that hydrophobic residues in σ_2 then bind and stabilize the exposed bases thereby initiating bubble formation. This initial melting of the DNA creates flexibility in the DNA, which allows the region downstream of −10 to bend and enter the primary channel. Interactions between the DNA and the active site then result in the further melting of the DNA, extending the transcription bubble past the transcription start site to form the definitive open complex (Plate 2.4C, Figure 2.9B).

3 **Abortive initiation.** Nucleotides enter through the secondary channel and transcription initiates. At this stage, σ is still bound to the −10 and −35 elements. Once the nascent transcript exceeds a few nucleotides in length, it collides with the linker between σ_3 and σ_4 that runs beneath the β-flap and occupies the RNA exit channel. This sets up a competition between the σ linker and the growing transcript for the same space. The nascent transcript often loses the competition resulting in the release of an abortive transcript.

4 **Promoter clearance.** Occasionally, the 5′ end of the nascent transcript is able to displace the σ_3–σ_4 linker from the channel beneath the flap.

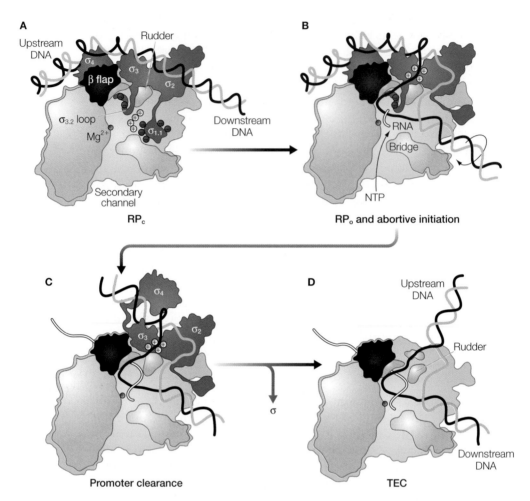

Figure 2.9 *Structural view of the steps in initiation.* The drawings depict schematic "cut-away" views into the active site of the bacterial RNA polymerase. This view can be generated, by rotating the structure shown in Plate 2.4C by about a third of a turn in the clockwise direction around an axis perpendicular to the page. In the closed complex (A), the DNA lies across the surface of the polymerase and is completely external to the active site. The σ_4 domain is bound to the tip of the β-flap and in turn binds the -35 element. The σ_2 domain binds the -10 element. In the open complex (B), melting of the DNA, favored by contacts between σ_2 and the exposed bases in the melted non-template strand -10 region, allows the DNA to bend and enter the active site through the primary channel. Transcription most often leads to abortive initiation because steric interference from the σ_3–σ_4 linker (also known as the $\sigma_{3.2}$ loop) prevents the 5' end of the nascent transcript from exiting the active site through the exit channel beneath the β-flap. Occasionally, however, the transcript displaces the linker allowing promoter clearance to occur (C). (D) The displacement of the σ_3–σ_4 linker destabilizes the association of the σ factor with the core polymerase, resulting in σ release and formation of the definitive ternary elongation complex (TEC). From Murakami, K.S. and Darst, S.A. (2003) *Curr Opin Struct Biol*, **13**, 31–39, with permission from Elsevier.

When the nascent transcript is ~10 nucleotides in length it is long enough to fully displace the linker from the exit channel (Figure 2.9C). This destabilizes the association between σ and the core polymerase, and the conformational change in σ also destabilizes the association between σ and the promoter. The resulting release of σ from the core polymerase and the promoter allows promoter escape (Figure 2.9D).

For some promoters, there may be, on average, hundreds of abortive initiation cycles before promoter clearance is achieved. What is the use of this seemingly wasteful scheme? RNA polymerases must be able to synthesize nucleic acid strands *de novo*. In other words, they require the ability to start with one nucleotide, which is added to a second nucleotide to create a dinucleotide, which is added to a third nucleotide to create a trinucleotide, and so on. This contrasts with DNA polymerases, which can normally only elongate pre-existing nucleic acids of at least 10 or so nucleotides in length. At the first step in *de novo* RNA synthesis, it is important that the first nucleotide be held tightly in the active site so that it can be positioned properly to make the attack on the second nucleotide. Inspection of the holoenzyme structure suggests that part of the tight binding pocket for this first nucleotide is provided by the σ_3–σ_4 linker in the RNA exit channel. Once the RNA strand begins to elongate, however, the pairing between the nascent RNA and the template DNA is sufficient to position the 3′ end of the nascent RNA for nucleophilic attack. At this point, the σ_3–σ_4 linker simply needs to get out of the way so that the 5′ end of the nascent RNA has somewhere to go as it exits the transcription bubble.

> **Nascent RNA** – literally means an RNA molecule being born. The transcript found in the ternary elongation complex is a nascent RNA.

Thus, the active site must be able to accommodate conflicting needs – tight packing is required during the early cycles of transcription to properly position the substrates, but a channel is required during later cycles of transcription to allow room for the nascent RNA to exit the active site. Apparently, these conflicting requirements are met by plugging the exit channel with the σ linker and then forcing the nascent RNA to push the plug out of the way.

The σ_1 domain, which due to disorder is invisible in the holoenzyme crystal structure, may also play a dynamic role in the process of initiation. In particular, σ_1 is thought to reside inside the active site cleft prior to open complex formation where it may serve to hold the channel open wide enough to allow the entry of DNA into the channel. After binding of the DNA in the channel, σ_1 may then exit the channel allowing the channel to clamp down on the DNA stabilizing the open complex and contributing to enzyme processivity (compare Figure 2.9A, B).

2.5 TRANSCRIPTIONAL TERMINATION

2.5.1 Why terminate?

Efficient transcription requires processivity, which in turn depends on the stability of the TEC. However, once the end of a gene is reached, it is necessary to trigger the break up of this stable complex. This is important for a number of reasons: first, continued transcription beyond the end of the gene would result in the wasteful consumption of nucleotides; second, RNA polymerases from one gene elongating through the promoter region of a downstream gene could interfere with the properly regulated expression of the downstream gene; and third, termination of transcription is required for the release and recycling of the core polymerase.

At least three interactions, all of which have already been discussed, contribute to the stability of the TEC. First, the association of the DNA template with the TEC is stabilized by the clamping of the crab claws around the template downstream of the transcription bubble. Second, the association of the nascent transcript with the TEC is stabilized by the 8–9 bp heteroduplex between the transcript and the template strand in the transcription bubble. Third, the binding of the transcript to the TEC is also stabilized by interactions between elements of the protein (the so-called RNA binding site, which includes the β′ rudder) and the transcript as it exits the transcription bubble. For transcription to terminate all these interactions must dissolve.

2.5.2 Termination in bacteria

Roughly half the genes in the *E. coli* genome undergo termination by a mechanism termed intrinsic termination, which is so named because it depends only upon interactions between the core polymerase and intrinsic termination signals in the transcript – no accessory factors are required. Other genes require accessory termination factors, which will not be discussed.

At the level of the DNA, the intrinsic termination signal consists of a GC-rich inverted repeat followed by a run of A residues in the template strand (Figure 2.10). Transcription of the A-rich region leads, by an unknown mechanism, to pausing of the polymerase giving the just synthesized GC-rich inverted repeat in the transcript time to snap back on itself and form a hairpin. This hairpin cannot coexist with the adjacent RNA–DNA heteroduplex. In the close confines of the active site and exit channel there is simply not enough room for both duplexes to form simultaneously. Because dA/U duplexes are of unusually low stability, the hairpin forms at the expense of the heteroduplex. The loss of these stabilizing basepairs between the transcript and the template weakens the

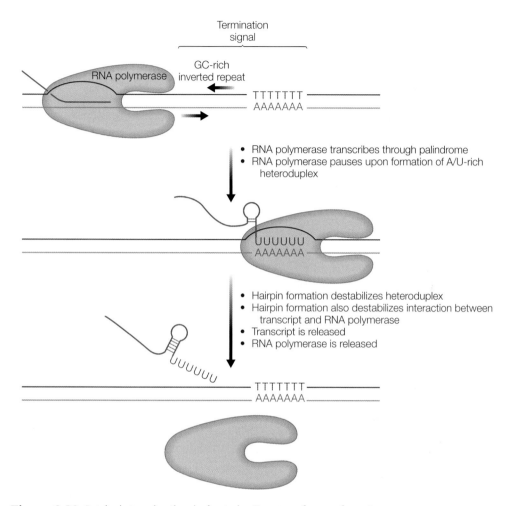

Figure 2.10 *Intrinsic termination in bacteria.* See text for explanation.

association of the transcript with the TEC. Hairpin formation is also thought to induce the loss of the stabilizing contacts between the transcript and the template and between the template and the protein leading to disassembly of the TEC.

2.5.3 Termination in eukaryotes

Class I and class III gene termination

Class I and class III termination in eukaryotes are partially reminiscent of intrinsic termination in bacteria. In both cases, termination occurs at discrete sites at the 3′ ends of genes and is triggered by the transcription of an A-rich stretch in the template strand. In these cases, however, there is no hairpin upstream of the A-rich termination signal. Instead, it appears that various mechanisms exist to induce pausing at these A-rich sequences, with the intrinsic instability of the dA/U heteroduplex then triggering the release of the transcript from the paused TEC.

In the case of class III genes, efficient termination seems to require the backtracking process discussed above in the section on elongation. Recall that when Pol II encounters a roadblock, backtracking often occurs and the resulting backtracked complex is then resolved when TFIIS stimulates hydrolytic cleavage of the backtracked RNA. In the case of Pol III, hydrolysis of backtracked complexes does not require TFIIS. Instead, the needed stimulatory activity is provided by the Pol III subunit Rpc11, which has a domain with homology to TFIIS. (The subunit in Pol II with homology to Rpc11 (Rpb9) lacks the TFIIS-like domain.)

During termination of Pol III at an A-rich termination signal in the template, the intrinsic instability of the dA/U-rich heteroduplex leads to backtracking. The Rpc11-stimulated hydrolysis of the backtracked complex then allows elongation to resume. It is thought that repeated rounds of backtracking and cleavage may result in increased time spent at the terminator, thereby allowing more time for transcript release.

Thus, backtracking-induced cleavage has distinctly different roles in class II and class III genes. In class II genes, this process is used to help the polymerase elongate through roadblocks, while in class III genes, it is used to increase the length of time spent at termination signals thereby increasing the probability of termination. This difference could be related to the different size of the class II and class III transcription units; the former can be many kilobases in length, while the latter are on the order of 100 bp in length. The probability of encountering a roadblock during the production of a functional transcript is therefore much greater for RNA polymerase II and mechanisms to deal with these road blocks are correspondingly more important. In the case of class III genes, where large numbers of small transcripts are produced, it may be more important to ensure efficient utilization of discrete termination signals.

Class II gene termination

In the case of class II genes, the 3' end of the mature transcript (mRNA) is defined not by the position of termination, but by a process known as cleavage and polyadenylation. In this process, a highly conserved AAUAAA signal in the transcript together with a downstream GU-rich sequence are recognized by a series of factors that direct the endonucleolytic cleavage of the transcript followed by the addition of several hundred adenylate residues to the 3' end of the transcript (see Chapter 1). The polymerase continues to transcribe downstream past the polyadenylation signal and gradually terminates transcription over a region of several hundred basepairs.

Termination by Pol II depends on the polyadenylation signal in the transcript itself and also on the cleavage and polyadenylation machinery. Thus, mutations in the AAUAAA signal that prevent processing also tend to block termination. Transcription through the polyadenylation signal results in the recruitment of a factor termed Pcf11 to the TEC, perhaps via interactions between Pcf11 and the polyadenylation signal in the transcript or between Pcf11 and RNA processing factors that bind the polyadenylation signal. Pcf11 then promotes the dissolution of the TEC by unknown mechanisms. Note that the above-described scheme ensures that termination will only occur after transcription has proceeded past a functional polyadenylation signal. This makes sense since polyadenylation is essential for mRNA function.

2.6 SUMMARY

We have seen that the transcription cycle contains distinct initiation, elongation, and termination phases, each of which can be subdivided into multiple subphases. Transcription in all cellular organisms is catalyzed by multisubunit core RNA polymerases, which are highly conserved in all three domains of life. These core polymerases are not sufficient for initiation, which requires additional factors. While bacteria and archaea each possess a single core polymerase, eukaryotes contain three core polymerases, which are responsible for transcribing different sets of eukaryotic genes. The subunit structure of the minimal catalytically active bacterial core polymerase is $\alpha_2\beta\beta'$. While the archaeal and eukaryotic core polymerases contain homologs of each of the bacterial subunits, they also contain many additional subunits of generally unknown function not found in the bacterial enzyme.

The structure of the core polymerase, which consists of two crab claws flanking an active site cleft, suggests how RNA polymerase can function as an intricate molecular machine to direct the accurate and processive synthesis of RNA. In the elongation complex, DNA reaches the active site through a primary channel between the crab claws. The DNA is denatured in the active site to form a transcription bubble. The growing RNA strand forms a heteroduplex with the template strand in this DNA bubble. At each nucleotide addition cycle, the hydroxyl group at the 3' end of the RNA strand attacks the α-phosphorus atom of an NTP, forming a new phosphoester linkage. Critical metal ions in the active site stabilize the transition state in this reaction thereby catalyzing the reaction. Nucleotides reach the active site through a pore known as the secondary channel. As the nascent transcript exits the polymerase, it travels underneath the β-flap. When the polymerase encounters roadblocks, it can backtrack. Reversion of the backtracked complex to a productive TEC requires factors that induce the polymerase to cleave the transcript generating a new 3' end in the active site.

In bacteria, initiation requires a σ factor, which binds the core polymerase to form the holoenzyme, recognizes the −10 and −35 elements in the promoter, and orchestrates the initiation sequence. This sequence includes closed complex formation, open complex formation, abortive initiation, and promoter clearance leading to formation of the stable TEC. Promoter clearance can only occur when the 5′ end of the transcript displaces a portion of σ factor from the space underneath the β-flap, thereby allowing the nascent transcript to exit the active site.

In bacteria, about half the genes are terminated by a process intrinsic to the core polymerase in which formation of a hairpin in the nascent transcript destabilizes the association of the transcript with the template. In eukaryotes, termination by Pol I and Pol III occurs by mechanisms akin to the intrinsic mechanism in bacteria. Termination by Pol II requires accessory factors that serve to coordinate termination with 3′ end processing.

PROBLEMS

1 The RNA polymerase active site can catalyze the hydrolysis of phosphoester bonds. How is the RNA polymerase active site designed to ensure that wasteful hydrolysis will not normally occur?

2 In DNA polymerases, editing is achieved through an exonucleolytic active site, which is completely separate from the polymerization active site. Errors in the newly synthesized DNA lead to the movement of the 3′ end of the nascent strand into the exonuclease active site, where the erroneous nucleotides are removed before the 3′ end moves back into the polymerization active site. How might errors near the 3′ end of a nascent transcript lead to editing by RNA polymerases, which lack a separate editing active site?

3 In addition to reducing the ability of nucleotides to reach the active site, microcin J25 might also be expected to have a deleterious effect on the ability of RNA polymerase to negotiate roadblocks in the template. Why is this?

4 Why might there be a conflict between transcriptional fidelity during the early stages of transcription and the need for RNA polymerases to function in a highly processive manner? How has bacterial RNA polymerase resolved this conflict?

5 Using recombinant DNA methodology, it is possible to construct a bacterial RNA polymerase that lacks the β-flap. Transcriptional elongation is fairly normal with core polymerase lacking the flap. However, the flapless holoenzyme initiates transcription very inefficiently. Explain these observations based on the structure of the core polymerase and holoenzyme.

6 Inosine (I) is an analog of guanosine that can be incorporated by RNA polymerases opposite C residues in the template. However, I:C

basepairs are of lower stability than are G:C basepairs. Why might the inclusion of inosine in a transcription reaction result in reduced efficiency of intrinsic termination?

FURTHER READING

RNA polymerase and the transcription cycle

Elongation and the catalysis of phosphoester bond formation

Yin, H., Wang, M.D., Svoboda, K., Landick, R., Block, S.M. and Gelles, J. (1995) Transcription against an applied force. *Science*, **270**, 1653–1657. *A demonstration of the forceful nature with which RNA polymerase moves down the template.*

Zaychikov, E., Martin, E., Denissova, L., Kozlov, M., Markovtsov, V., Kashlev, M., Heumann, H., Nikiforov, V., Goldfarb, A. and Mustaev, A. (1996) Mapping of catalytic residues in the RNA polymerase active center. *Science*, **273**, 107–109. *Proof that the RNA polymerase catalytic site maps to a series of highly conserved aspartate residues.*

Initiation

Travers, A.A. and Burgess, R.R. (1969) Cyclic re-use of the RNA polymerase σ factor. *Nature*, **222**, 537–540. *The earliest demonstration that bacterial RNA polymerase holoenzyme consists of a core that catalyzes RNA synthesis and a σ factor required for initiation.*

Carpousis, A.J. and Gralla, J.D. (1985) Interaction of RNA polymerase with lacUV5 promoter DNA during mRNA initiation and elongation. Footprinting, methylation, and rifampicin-sensitivity changes accompanying transcription initiation. *J Mol Biol*, **183**, 165–177. *An elegant study demonstrating the basic features of open complex formation, abortive initiation, and promoter clearance.*

Termination

Gusarov, I. and Nudler, E. (1999) The mechanism of intrinsic transcription termination. *Mol Cell*, **3**, 495–504. *Evidence that the intrinsic termination signal functions to induce pausing followed by the dissolution of the contacts that stabilize the TEC.*

Chedin, S., Riva, M., Schultz, P., Sentenac, A. and Carles, C. (1998) The RNA cleavage activity of RNA polymerase III is mediated by an essential TFIIS-like subunit and is important for transcription termination. *Genes Dev*, **12**, 3857–3871. *A demonstration that the intrinsic ability of the RNA polymerase III to resolve backtracked complexes is needed for efficient termination.*

Zhang, Z. and Gilmour, D.S. (2006) Pcf11 is a termination factor in Drosophila that dismantles the elongation complex by bridging the CTD of RNA polymerase II to the nascent transcript. *Mol Cell*, **21**, 65–74. *A connection between the 3' end processing machinery and termination by RNA polymerase II.*

RNA polymerase structure and function

Bacterial RNA polymerases

Zhang, G., Campbell, E.A., Minakhin, L., Richter, C., Severinov, K. and Darst, S.A. (1999) Crystal structure of Thermus aquaticus core RNA polymerase at 3.3 A resolution. *Cell,* **98**, 811–824. *The first high resolution structure of a multi-subunit core RNA polymerase.*

Korzheva, N., Mustaev, A., Kozlov, M., Malhotra, A., Nikiforov, V., Goldfarb, A. and Darst, S.A. (2000) A structural model of transcription elongation. *Science,* **289**, 619–625. *A model for the bacterial TEC based on an extensive series of cross-linking studies.*

Vassylyev, D.G., Sekine, S., Laptenko, O., Lee, J., Vassylyeva, M.N., Borukhov, S. and Yokoyama, S. (2002) Crystal structure of a bacterial RNA polymerase holoenzyme at 2.6 A resolution. *Nature,* **417**, 712–719. *A high resolution structure of the bacterial RNA polymerase holoenzyme showing how the core interacts with σ factor.*

Murakami, K.S. and Darst, S.A. (2003) Bacterial RNA polymerases: the whole story. *Curr Opin Struct Biol,* **13**, 31–39. *A review article relating the structure of RNA polymerase to the steps in the transcription cycle.*

RNA polymerase II

Cramer, P., Bushnell, D.A. and Kornberg, R.D. (2001) Structural basis of transcription: RNA polymerase II at 2.8 angstrom resolution. *Science,* **292**, 1863–1876. *First high resolution structure of a eukaryotic RNA polymerase.*

Gnatt, A.L., Cramer, P., Fu, J., Bushnell, D.A. and Kornberg, R.D. (2001) Structural basis of transcription: an RNA polymerase II elongation complex at 3.3 A resolution. *Science,* **292**, 1876–1882. *A "snapshot" of the Pol II elongation complex.*

3

The eukaryotic basal machinery

Key concepts

- Eukaryotic and archaeal core RNA polymerases each require a different set of general transcription factors to recognize core promoters and initiate transcription
- All eukaryotic and archaeal polymerases require a universal general factor termed TBP – in the case of class II core promoters, TBP binds the DNA and nucleates a preinitiation complex
- Most eukaryotic and archaeal polymerases also require a general factor of the TFIIB family, which functions analogously to bacterial σ factors
- RNA polymerase II is unique among RNA polymerases in requiring an ATP-dependent helicase to open promoters. This helicase activity is supplied by TFIIH

3.1 INTRODUCTION

As we saw in the previous chapter, RNA polymerase is a multisubunit machine with a structure that is conserved in all three domains of life. In bacteria, the core enzyme contains four or five subunits that are sufficient for processive RNA synthesis; homologs of these subunits are also found in the more complex archaeal and eukaryotic enzymes. The bacterial RNA polymerase holoenzyme also includes one of multiple alternative σ subunits, which confer promoter specificity upon the core polymerase. In archaea and eukaryotes, however, there are no σ homologs. Instead, each RNA polymerase interacts with a set of general transcription factors, which directs the polymerase to the appropriate set of promoters and assists the polymerase in opening and clearing the promoter.

Each archaeal/eukaryotic core polymerase utilizes a distinct set of general transcription factors, although there is some overlap and structural homology between the sets (see Tables 3.1–3.3). The complexity of the general machinery seems to be at least roughly related to the complexity of the set of genes transcribed by the polymerase. Therefore, in eukaryotes, which contain thousands of class II genes that must be transcribed in intricately regulated spatial and temporal patterns, the set of general factors that interact with RNA polymerase II during initiation is extremely complex – consisting of dozens of polypeptides with a combined molecular mass of several million daltons.

Class II genes are by far the most numerous of the eukaryotic genes. They are also subject to much more intricate and intensive regulation than class I and class III genes. Consequently, an understanding of the mechanisms of initiation at class II promoters is essential to an understanding of gene regulation. Most of this chapter will therefore examine the function of the class II general machinery. Where appropriate, the other archaeal and eukaryotic transcription systems will be introduced for purposes of comparison.

3.2 THE CLASS II CORE PROMOTER

Eukaryotic protein-encoding genes often have extremely complex cis-regulatory regions, including multiple cis-regulatory modules (CRMs) (Figure 3.1). Each CRM is itself often a complex module consisting of multiple binding sites for numerous regulatory factors including activators and repressors. These modules can be located a long distance from the site of initiation (up to several hundred thousand basepairs away) and can be upstream or downstream of the initiation site. While CRMs are sometimes considered part of the "promoter", the term promoter is more correctly used to denote just the sequences immediately surrounding the site of initiation that are sufficient for specifically initiated transcription in a cell-free system. To remove any ambiguity, the sequences

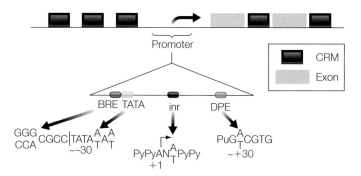

Figure 3.1 *Structure of a typical class II gene.* In metazoans most genes are split into two or more exons. Cis-regulatory modules (CRMs) are usually a few hundred basepairs in length and can be located upstream of the gene, downstream of the gene, or within introns. They can be several hundred thousand basepairs from the promoter. In budding yeast, which has a much more compact genome than metazoans, split genes are the exception rather than the rule and CRMs are usually located in the 5′ flanking region within several hundred basepairs of the promoter. Class II promoters are characterized by a number of conserved elements such as the TATA box, the B recognition element (BRE), the initiator (inr), and the downstream promoter element (DPE). Most promoters contain only one or two of these elements.

around the transcriptional start site are often referred to as the "core promoter". For class II genes, this region includes the transcriptional start site and no more than 50 bp upstream and 50 bp downstream of the start site. While eukaryotic core promoters are sometimes sufficient for specifically initiated transcription in a test tube, they are rarely sufficient for detectable levels of transcription in a living cell, which usually requires activation of the core promoter by one or more CRMs.

Class II core promoters vary tremendously from gene to gene; there is no simple way to predict if a segment of DNA will function as a promoter by inspecting its sequence. Nonetheless, alignment of large numbers of experimentally defined core promoters has revealed several conserved motifs (Figure 3.1). The best-characterized core promoter motif is the TATA box, which, in animals, is located about 30 bp upstream of the start site. The consensus sequence of the TATA box is very similar to that of the −10 box found in bacterial σ^{70} promoters. However, as will become apparent, the TATA box is recognized in a completely different way from the bacterial −10 box and therefore the resemblance is probably coincidental. The TATA box sometimes contains an upstream extension termed the B recognition element (BRE). Other motifs found in class II core promoters include the initiator (inr), which straddles the initiation site, and the downstream promoter element (DPE), which is found about 30 bp downstream of the initiation site (Figure 3.1). Promoters rarely contain all of these conserved motifs (the TATA box, the BRE, the inr, and the DPE) – usually they contain just one or two.

3.3 THE CATALOG OF FACTORS

Purified eukaryotic RNA polymerase II is a 12-subunit enzyme, with a total molecular mass of about 500 kD (see Chapter 2). Despite this complexity, it is incapable of initiating transcription in a promoter-specific manner. In the late 1970s, however, it was discovered that when RNA polymerase II was supplemented with a crude whole cell or nuclear protein extract, it could specifically initiate transcription from class II promoters. The subsequent fractionation of the crude extracts led to the identification and later the purification of a large set of factors that were required for promoter-specific initiation (Box 3.1): these are named transcription factors for RNA polymerase II A, B, D, E, F, and H, or more simply TFIIA, TFIIB, TFIID, TFIIE, TFIIF, and TFIIH (Table 3.1).

The class II general machinery is highly conserved in eukaryotic evolution; eukaryotes as diverse as yeast and mammals employ the same six general factors. At least some of the components of the class II machinery are of particularly ancient origin. For example, TATA-binding protein (TBP), a subunit of TFIID, is a component of the class I and class III machinery, as well as the class II machinery (Table 3.2). Thus, TBP is sometimes considered to be a "universal" general transcription factor, because it is required for transcription of all three classes of eukaryotic genes. TFIIB also appears to have arisen prior to the divergence of the three eukaryotic RNA polymerases since a TFIIB homolog termed B-related factor (BRF) is found in the class III machinery. Further attesting to the broad importance and early evolution of TBP and TFIIB, both of these factors have counterparts in archaea that are essential for all archaeal transcription (Table 3.2). In fact, the essential components of the archaeal general transcriptional apparatus are simply TBP, the archaeal TFIIB homolog termed TFB, and core RNA polymerase.

> **What is the universal function of TBP?**
>
> TBP is clearly of fundamental importance since it is required for all eukaryotic and archaeal transcription. However, the nature of the universally required function is a mystery. The universal function is clearly not TATA binding, since many class II and nearly all class I and class III promoters lack TATA boxes.

3.4 PATHWAY TO THE PREINITIATION COMPLEX

Transcriptional initiation by RNA polymerase II involves the assembly of a nucleoprotein complex termed the preinitiation complex (PIC) that

Box 3.1

Discovery of the general transcription factors

Although eukaryotic RNA polymerases were first purified in the 1960s, it was more than 10 years before biochemists learned how to coax these enzymes into initiating transcription at promoters rather than just at random nicks and gaps in DNA. The key was to take the purified polymerases and "unpurify" them, that is, add them back to crude protein extracts containing the general transcription factors. Although transcription is a nuclear process, the first cell-free transcription systems were, strangely enough, prepared from cytoplasmic extracts of cultured mammalian cells – apparently enough of the general factors leak out of the nuclei during cell lysis to allow the reconstitution of promoter-specific initiation when the cytoplasmic fraction is supplemented with purified RNA polymerases and programmed with appropriate promoters. Within a few years, however, biochemists learned how to prepare soluble protein extracts from the nuclei of HeLa cells (a human carcinoma cell line) that were a much richer source of the activities required for promoter-specific transcription.

One highly successful extraction procedure that is still used today involves isolating nuclei and suspending them in a medium containing 0.42 M NaCl (an empirically determined optimum). The process of breaking open cells damages and therefore permeabilizes the nuclear envelopes. The salt serves to disrupt the ionic bonds stabilizing macromolecular interactions inside the permeabilized nuclei, allowing many nuclear proteins to diffuse freely out of the nuclei into the medium. The chromatin is not extracted by this procedure, but is removed along with the nuclear envelopes in a subsequent ultracentrifuge spin. Interestingly, although these nuclear extracts contain some RNA polymerase II, much of the Pol II is not extracted by this procedure, but remains bound to the chromatin, probably a reflection of the great stability of the ternary elongation complex.

The purification of the general factors from the extracts involved years of grueling work by dozens of graduate students and postdoctoral fellows working long hours in the cold room to guard against thermal denaturation and proteolytic degradation of the factors. The approach was classical column chromatography and involved much trial and error (mostly error) to find conditions that would allow for the separation and eventual purification of all the general factors.

A favorite class II promoter for these purification efforts was the adenovirus major late promoter, which is highly active due to the presence of a consensus TATA box and a consensus initiator. Any chromatographic fraction that was required to reconstitute transcription of this promoter was deemed to contain one or more class II general transcription factors.

One chromatographic technique that was used to great advantage in the purification of general factors was phosphocellulose chromatography. Many of the factors required for transcription bind to phosphocellulose beads perhaps because the arrangement of phosphate groups along this carbohydrate polymer resembles the arrangement of phosphate groups along the DNA backbone. Once bound to phosphocellulose, factors can be eluted from the beads by gradually increasing the salt concentration in the medium. The cations in the salt solution bind the phosphate groups on the beads and displace the bound proteins. Different proteins bind the phosphocellulose with different avidity and are therefore released at different salt concentrations. This approach allowed the resolution of three fractions that were required to reconstitute transcription of the 5s rRNA promoter by Pol III (TFIIIA, TFIIIB, and TFIIIC) as well as three fractions that were required to reconstitute transcription of the adenovirus major late promoter by Pol II (TFIIA, TFIIB, and TFIID) (Figure B3.1). Additional class II factors were later found by the further fractionation of the phosphocellulose fractions. For example, the fraction containing TFIIB also contained TFIIE, TFIIF, and TFIIH, which were eventually separated from one another by additional column chromatography steps.

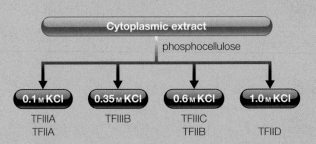

Figure B3.1 *Scheme used in the early phosphocellulose fractionation of class II and class III general transcription factors.* Cytoplasmic extracts were applied to a phosphocellulose column, which was then washed with the indicated concentrations of KCl to elute the indicated transcription factors. A very similar approach can be used to fractionate nuclear extracts.

contains RNA polymerase, all six general transcription factors, and the core promoter. PIC assembly begins with the recognition of the promoter by TFIID, an event that is facilitated by TFIIA (Figure 3.2). As mentioned above, TFIID contains a subunit called TBP and the interaction between TBP and the TATA box is critical in the initial step of PIC assembly at TATA box-containing promoters. Interestingly, however, even TATA-less promoters (constituting at least half of the class II promoters) require TBP. In the case of TATA-less promoters, recognition may involve interactions between other subunits of TFIID (the so-called TBP-associated factors or TAFs, see below) and the inr or DPE. Once TFIID is bound, it recognizes and recruits TFIIB to the complex. TFIIB in turn recruits a preformed complex of TFIIF and RNA polymerase II. Finally, TFIIE and TFIIH bind, completing formation of the PIC.

Table 3.1

The class II general transcription factors

Factor	Subunit names and molecular masses (kD)	Functions
TFIIA	α (37) β (19) γ (13)	• Stabilizes association of TFIID with promoter
TFIIB[1]	Single subunit (35)	• Helps recognize promoter through sequence-specific contacts between helix-turn-helix motif in TFIIB$_C$ and BRE • Non-specific contacts between TFIIB$_C$ and DNA downstream of TATA box stabilize PIC • Zinc ribbon, B-finger, and TFIIB$_C$ all bind Pol II completing the bridge from TBP to Pol II • Zinc ribbon domain binds Pol II dock • B-finger extends through RNA exit channel into active site and mediates start site selection
TFIID	TBP (38) 10–15 TAFs	• TBP recognizes and binds the TATA box • TBP also required for PIC formation at TATA-less promoters • See Table 3.3 for TAF functions
TFIIE[2]	α (56) β (34)	• Binds DNA non-specifically • Binds to Pol II and helps recruit TFIIH
TFIIF[2]	RAP74 (58) RAP30 (26)	• Associates with Pol II • Contacts TFIIB extensively in core, zinc ribbon, and B-finger domains and stabilizes PIC • Binds DNA non-specifically and promotes wrapping of promoter DNA around the PIC • Assists TFIIB in start site selection • Stimulates elongation

TFIIH	Core TFIIH
	  XPB (89)
	  XPD (80)
	  p62 (62)
	  p52 (52)
	  p44 (44)
	  p34 (34)
	  p8 (8)
	Cyclin-activating kinase (CAK)
	  CDK7 (40)
	  cyclin H (37)
	  MAT-1 (32)
		• Core TFIIH is an essential component of nucleotide excision repair machinery
		• 3′–5′ helicase activity of XPB is essential for open complex formation
		• 5′–3′ helicase activity of XPD is dispensable for transcription
		• Phosphorylation of Pol II CTD (Ser5) by CAK is associated with promoter clearance and recruitment of the capping enzyme

BRE, B recognition element; CAK, cyclin-activating kinase; CTD, C-terminal domain; PIC, preinitiation complex; TBP, TATA-binding protein.

[1]TFIIB consists of a C-terminal domain (TFIIB$_C$) and an N-terminal domain (TFIIB$_N$). The N-terminal domain contains a zinc ribbon motif followed by a B-finger.

[2]Both TFIIE and TFIIF consist of two distinct gene products. However, TFIIF is known to form a tetramer in solution and both TFIIE and TFIIF may enter the PIC as tetramers containing two molecules of each polypeptide.

Table 3.2

Human class I, human class III, and archaeal (*Pyrococcus woesei*) general transcription factors compared[1]

Class I		Class III[2]		*P. woesei*[3]	
Factor	Subunit names and lengths (residues)	Factor	Subunit names and lengths (residues)	Factor	Subunit names and lengths (residues)
SL1	**TBP (339)** TAF1A (336/450)[4] TAF1B (588) TAF1C (869/775)[4]	TFIIIB	**TBP (339)** *BRF (675)*[2] B″ (1388)[2]	**TBP**	Single polypeptide (191)
Ubf1	Single polypeptide (764)	TFIIIC[5]	C220 (2140) C110 (911) C102 (886) C90 (822) C63 (519)	*TFB*	Single polypeptide (300)

[1]Values in parentheses are lengths in amino acid residues. Each set of factors contains TATA-binding protein (TBP; in bold), which is therefore a "universal" transcription factor. Every set of factors except the class I factors contain a TFIIB homolog (in italics).
[2]BRF (B-related factor) and B″ are, formally speaking, TBP-associated factors, although the TBP-associated factor (TAF) nomenclature is not generally employed for them.
[3]Archaeal genomes such as those of *P. woesei* encode a protein with homology to TFIIE α subunit termed TFE. While TFE is stimulatory, it is not essential for transcription.
[4]TAF1A and TAF1C each exist in two isoforms encoded by alternative splice variants. The lengths of the two isoforms are given.
[5]While the vast majority of class III promoters (tRNA promoters and 5s rRNA promoters, require TFIIIC) there are a few such promoters (e.g., the human U6 snRNA promoter) that do not require TFIIIC. Therefore, strictly speaking TFIIIB is the only completely general class III factor.

The PIC is analogous to the closed complex defined in studies of bacterial transcription. As in bacteria, the formation of this closed complex is followed by promoter opening and promoter clearance steps. Some of the general transcription factors, especially TFIIH, play important roles in these steps.

The classical assembly pathway outlined above is one of multiple allowable pathways to the PIC. For example, under some conditions many of the general factors may preassociate with one another and RNA polymerase in solution to form a so-called "Pol II holoenzyme", which could conceivably join the PIC in a single concerted step. While the classical assembly pathway is not the unique pathway, its characterization played an essential role in helping us to understand many of the complexities of PIC function to be discussed below.

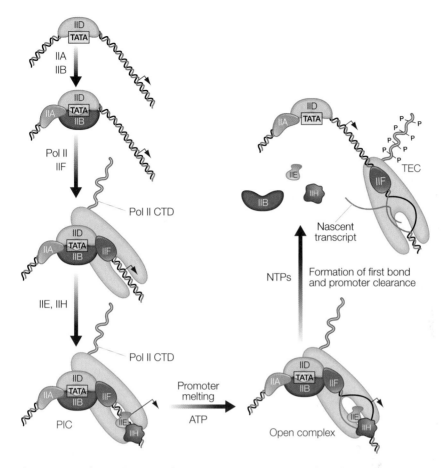

Figure 3.2 *The initiation pathway at class II promoters.* The classical assembly pathway for a TATA box-containing promoter is shown. The primary event in recognition of the promoter is the binding of TBP (a subunit of TFIID) to the TATA box. TATA-less promoters must be recognized in a different way, probably involving interactions between TBP associated factors in TFIID and other core promoter elements such as the inr and the DPE. Recognition of promoters is probably also dependent, in some cases, upon activator proteins bound to promoter proximal CRMs, which then recruit components of the general machinery. The classical pathway for preinitiation complex (PIC) assembly shown here is only one of multiple possible pathways for PIC assembly. For example, in some cases, Pol II may preassemble with TFIIB, TFIIF, TFIIE, and TFIIH, which then join the PIC in a single concerted step. In addition, promoters have been characterized in which binding of TFIID actually occurs after binding of Pol II. When Pol II joins the promoter, the C-terminal domain (CTD) is in an unphosphorylated state. This favors association of the Mediator, which plays an important role in activation (see Chapter 4). Promoter clearance is accompanied by extensive phosphorylation of the CTD, primarily on Ser5, leading to release of the Mediator. TFIID and TFIIA may stay behind at the promoter after initiation where they can nucleate another PIC, while TFIIB, TFIIE, and TFIIH are released. TFIIF moves down the template with the elongating polymerase and appears to be important for elongation efficiency. NTP, nucleoside triphosphate; TEC, ternary elongation complex.

3.5 PROMOTER RECOGNITION AND NUCLEATION OF THE PIC BY TFIID

3.5.1 TATA box recognition by TBP

TBP, the component of TFIID directly responsible for recognition of the TATA box, was initially purified from yeast extracts as a polypeptide capable of substituting for the TFIID fraction in a HeLa cell transcription system. The finding that TBP could substitute for TFIID led to the conclusion that TFIID and TBP were equivalent to one another. As will be discussed below, this conclusion was later found to be erroneous – in reality, TFIID is a multisubunit protein, of which TBP is but one subunit.

HeLa cells – a human cervical cancer cell line.

TBP contains a variable N-terminal domain that is dispensable for transcription and an essential highly conserved C-terminal domain of about 180 amino acids in length, which consists of two degenerate repeats of an approximately 90 amino acid sequence (Plate 3.1A). Each repeat assumes the same basic fold, the major feature of which is a five-stranded antiparallel β-sheet (Plate 3.1B–E). These two subdomains come together forming a symmetric 10-stranded antiparallel β-sheet that has the overall shape of a saddle (Plate 3.1B, C).

Binding of TBP to the TATA box dramatically distorts the structure of the double helix to allow an induced fit between the protein and the DNA. Specifically, the protein splays open the minor groove and unwinds the DNA allowing the β-sheet to make extensive contact with the minor groove over the length of the TATA box. In addition, the protein induces a severe bend in the DNA, with the direction of the DNA changing by about 90° over the course of the TATA box (Plate 3.1E). Most of this bend comes at two sites in the DNA where phenylalanine residues in loops between β-strands (the "stirrups" of the saddle) are inserted between successive basepairs, wedging them apart (Plate 3.1C).

Major and minor grooves – the depressions between the sugar–phosphate backbones in double-stranded DNA through which the edges of the bases are visible. The grooves on opposite sides of the helix are of different widths. The wider of the two is termed the major groove, while the other is termed the minor groove.

What is the functional significance of this highly unusual structure, which is unlike that of any other known protein–DNA complex? The answer to this question is unclear. Perhaps by distorting the DNA, TBP destabilizes the DNA duplex thereby facilitating promoter opening. In addition, DNA distortion may help the transcriptional machinery contend with chromosomal proteins such as histones. These proteins organize the DNA in eukaryotic cells into a highly regular

and compactly folded nucleoprotein complex called chromatin (chromatin structure is discussed in detail in Chapter 5).
Chromatin formation has a repressive
effect on transcription by masking pro-

> **Histones** – small positively charged
> proteins that are major components of
> chromatin.

moter DNA from the transcriptional machinery. The distorted DNA
structure induced by TBP is incompatible with normal chromatin structure
and may therefore facilitate PIC formation by unmasking the promoter.

3.5.2 TBP-associated factors and their function in TFIID

In eukaryotic cells, most TBP is found in several protein complexes. The
polypeptides in these complexes are termed TBP-associated factors or TAFs.
As mentioned previously, TBP is a universal transcription factor in that
it is a component of the general transcription machinery for all three
eukaryotic RNA polymerases as well as the archaeal polymerase. TBP associates with three sets of TAFs to form three general transcription factors,
one for each eukaryotic RNA polymerase – TFIID for Pol II, SL1 for Pol I,
and TFIIIB for Pol III (see Tables 3.1 and 3.2). TFIID, which is a complex
of TBP and ~14 TAFs (Table 3.3), exhibits a three-lobed horseshoe-like
structure (Figure 3.3). TBP is located in or near the central lobe of the
horseshoe on its concave side (Figure 3.3B) suggesting that the promoter
may run through the central channel of the horseshoe.

One notable feature of the class II TAFs is the sequence and structural
similarity between many of the TAFs and core histones, which are the
major protein components of chromatin. These histones associate with

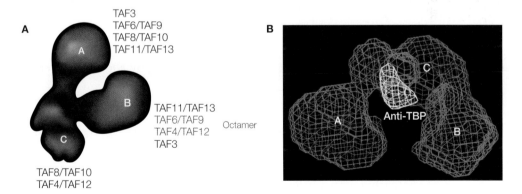

Figure 3.3 *Structure of TFIID as determined by electron microscopy.* (A) Yeast TFIID exhibits
a trilobular structure. Many of the TBP-associated factors (TAFs) have been mapped
to the various lobes by labeling the complex with antibodies against the TAFs. The
locations of these TAFs are indicated. (B) Electron microscopy reconstruction of human
TFIID exhibits a similar trilobular structure. The position of TBP, as determined by
immunolabeling, is indicated. B, from Andel, F., 3rd et al. (1999) *Science*, 286,
2153–2156, with permission from AAAS.

Table 3.3

Class II TAFs

Unified nomenclature[1]	Mass of human ortholog/mass of yeast ortholog (kD)[2]	Domains/catalytic functions	Contacts made in PIC
TAF1	250/145	Two bromodomains Histone acetyltransferase activity Serine/threonine kinase activity Ubiquitin conjugating activity	Inr
TAF2	150/150		Inr
TAF3	140/47	HFD	
TAF4:TAF12[3]	135/48:20/61	HFD	
TAF5	100/90	WD repeat domain	TFIIF
TAF6:TAF9[3]	80/60:32/17	HFD	DPE (TAF6) TFIIB (TAF9)
TAF7	55/67		
TAF8:TAF10	43/65:30/25	HFD	
TAF11:TAF13	28/40:18/19	HFD	
TAF14	$-^4$/30		
TAF15	68/$-^4$		

DPE, downstream promoter element; HFD, histone fold domain; PIC, preinitiation complex; TAF, TBP-associated factor.

[1]When two HFD containing TAFs are known to form a histone fold pair, they are listed together in the same cell of the table and separated by a colon.

[2]The masses of the human and yeast orthologs are separated by a slash, with the mass of the human ortholog listed first. Masses of the histone fold pair partners are separated by a colon.

[3]The TAF4:TAF12 and TAF6:TAF9 heterodimers can associate to form a histone octamer-like particle containing two copies of each heterodimer.

[4]No human ortholog has been reported for TAF14, while no yeast ortholog has been reported for TAF15.

one another to form heterodimers, and the formation of these dimers is mediated by a domain in each histone termed the histone fold domain (HFD). At least half of the class II TAFs contain HFDs (Table 3.3). These are essential for the stability of TFIID, and may provide protein interaction surfaces that help to solidify the complex.

The TAFs in TFIID have multiple functions. TAF1, for example, has several catalytic domains that direct a number of post-translational protein modifications including phosphorylation and acetylation. Although the physiological targets of these TAF1-catalyzed modifications are not well defined, the modification of transcription factors or histones by TAF1 may promote steps important in assembly or activation of the PIC. In addition, TAF1 and TAF2 may contact the inr, while TAF6 contacts the DPE thus explaining how the class II machinery can recognize TATA-less promoters.

Finally, TAFs have roles in transcriptional activation. When TBP was originally purified, it was found to substitute for TFIID in a basal cell-free transcription system, and therefore it was thought that TBP and TFIID were equivalent. However, it was soon discovered that a variety of transcriptional activators could not stimulate transcription in cell-free systems containing TBP in place of TFIID. It was this observation that led to the realization that TFIID must contain additional subunits required for activation and thus to the discovery of the class II TAFs. This role in activation partly reflects a role for TAFs as adaptors that help mediate the interactions between activators and the general machinery.

3.6 TFIIB: A FUNCTIONAL ANALOG OF BACTERIAL σ FACTORS

As mentioned above, eukaryotes and archaea do not contain σ factors. However, many of the functions of the bacterial σ factors seem to be carried out in eukaryotes and archaea by TFIIB family proteins.

3.6.1 TFIIB as a bridge between the promoter and polymerase

In the classical PIC assembly pathway, TFIIB binds immediately after TFIID and immediately before Pol II/TFIIF (Figure 3.2). This implies that TFIIB serves as a bridge between TFIID and Pol II. Structural and biochemical studies on promoter/TBP/TFIIB and RNA polymerase/TFIIB complexes confirm the existence of this bridge and illustrate its features.

TFIIB C-terminal domain and promoter directionality

TFIIB contains N- and C-terminal domains referred to as $TFIIB_N$ and $TFIIB_C$, respectively (Plate 3.1A). $TFIIB_C$ contains two direct repeats, which fold into homologous subdomains, each consisting of a bundle of five α-helices (Plate 3.1D, E). In the promoter/TBP/TFIIB complex, the C-terminal stirrup of TBP is inserted into the cleft between the two subdomains of $TFIIB_C$ (Plate 3.1E). In essence, the two $TFIIB_C$ subdomains form a clamp around the TBP stirrup.

In addition to contacting TBP, $TFIIB_C$ makes intimate sequence-specific contact with the DNA upstream of the TATA box. Two adjacent α-helices in $TFIIB_C$ fold into a so-called helix-turn-helix motif. This motif will be discussed in detail in Chapter 4. Suffice it to say for now that many sequence-specific DNA-binding proteins use such a motif to recognize and bind to specific DNA sequences. In $TFIIB_C$, the helix-turn-helix motif binds to a region termed the BRE just upstream of the TATA box, the sequence of which is conserved in many class II promoters (Plate 3.1D).

The sequence-specific interaction between TFIIB and the BRE helps to impose directionality on the promoter. Naturally occurring TATA boxes are unidirectional elements – transcription invariably leads in one direction from the TATA box. However, the structural basis for this unidirectionality is not obvious since the TATA box is partially symmetric (i.e., its sequence and that of its reverse complement are very similar to one another). As a result, TBP is able to bind TATA boxes in either orientation. Significantly, however, $TFIIB_C$ can only form a clamp around the C-terminal stirrup and not the N-terminal stirrup of TBP because of minor structural differences in the two stirrups. The binding of the $TFIIB_C$ helix-turn-helix motif to the BRE in turn imposes an asymmetry upon the TBP/TATA box complex, forcing the C-terminal repeat in TBP to bind the half of the TATA box adjacent to the BRE. The resulting directionality of the promoter/TBP/TFIIB complex imposes directionality upon the entire PIC.

An active role for TFIIB in start site selection

In addition to contacting TBP, TFIIB also contacts Pol II thus forming a bridge between TBP and RNA polymerase. The TFIIB N-terminal domain contains a zinc-binding motif (the zinc ribbon) followed by a protein loop (the B-finger), both of which are thought to make important contacts with Pol II (Plate 3.1A). A model of the Pol II/TBP/TFIIB/promoter complex (Plate 3.2) suggests that the zinc ribbon binds close to the exterior opening of the RNA exit channel. From there, the TFIIB polypeptide chain is thought to thread into the active site through the RNA polymerase exit channel placing the B-finger in the active site cleft. Mutations in the B-finger can alter the transcriptional start site, suggesting that the B-finger has an active role in positioning the DNA in the active site (Box 3.2).

Note that the position of a portion of TFIIB in the RNA exit channel in the Pol II PIC is equivalent to the position of the σ_3–σ_4 linker in the RNA exit channel of bacterial polymerase (see Chapter 2). Therefore, just as formation of the stable ternary elongation complex (TEC) requires the release of σ from the core polymerase in bacteria, class II TEC formation may require the release of TFIIB.

3.6.2 BRF, a TFIIB paralog in the class III machinery, and promoter melting

TFIIB family proteins have also been implicated in promoter melting leading to open complex formation, perhaps best illustrated by studies of class III promoters. The most numerous group of class III promoters are the tRNA promoters – often taken as paradigms for all class III promoters. Two general transcription factors, TFIIIB and TFIIIC, are required for promoter recognition and initiation at tRNA genes.

Box 3.2

Genetic screen for mutations that alter the start site

The application of genetic approaches to the study of transcription in the budding yeast *Saccharomyces cerevisiae* has provided much useful information about the workings of the basic transcriptional machinery. For example, genetic analysis of TFIIB reveals that this factor is more than a passive tether between TBP and RNA polymerase. Rather TFIIB may play an active role in orchestrating the initiation process.

In budding yeast, the gene encoding TFIIB was first identified in a screen for second site suppressors of a mutation in cyc1, the gene encoding cytochrome c1. The particular cyc1 allele used in these experiments has an aberrant ATG start codon upstream of the normal cyc1 start codon (Figure B3.2A). Because this

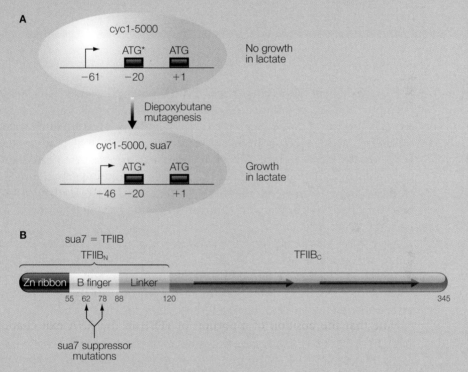

Figure B3.2 *Identification of TFIIB mutations that alter the transcriptional start site.* (A) Genetic screen for suppressors of cyc1. In this map of cyc1, the position of the normal ATG start codon is denoted as +1. The normal transcriptional start site is at −61. The cyc1-5000 mutant allele contains an aberrant start codon at −20 (marked with an asterisk) that interferes with translation. The cyc1-5000 cells are therefore unable to grow in a non-fermentable carbon source such as lactate. The cyc1-5000 strain was mutagenized and the mutagenized cells were grown in lactate medium to select for second site suppressor mutations. The genes defined by these mutations were denoted SUA1 to SUA8. The sua7 suppressor mutations alter the transcriptional start site to position −46. (B) SUA7 encodes TFIIB. The bar presents a schematic representation of the TFIIB domain structure. The two sua7 mutations isolated in the suppressor screen map to amino acids 62 and 78 in the B-finger domain, implicating this domain in start site selection.

aberrant start codon is not in the same reading frame as the cyc1 coding region, translation from the aberrant start codon results in a non-functional gene product. Since cytochrome c1 is essential for respiration, cells containing this mutant cyc1 allele fail to grow on a non-fermentable carbon source such as lactate. A screen for second site suppressors of this cyc1 allele turned up mutations in a number of genes, which were named suppressor of upstream ATG (SUA) genes. Two of these mutations defined the SUA7 gene, which was later discovered to encode TFIIB. These mutations both altered amino acids in the B-finger region of TFIIB and suppressed the mutant phenotype by shifting the transcriptional start site to a position downstream of the aberrant ATG. These findings strongly suggested an active role for TFIIB in the initiation process and, in particular, a role for the B-finger in the selection of the transcriptional start site.

PIC assembly at tRNA genes begins with the recognition of downstream promoter elements by TFIIIC (Figure 3.4). tRNA genes contain two conserved elements in the coding region (box A and box B) required for recognition by TFIIIC. Once bound, TFIIIC recruits TFIIIB, which serves as a bridge to RNA polymerase III. tRNA promoters were the first eukaryotic promoters to be defined, and the discovery that the only required sequence elements were downstream of the transcriptional start site was quite a surprise because it was in dramatic contrast with what had been learned about bacterial promoters.

As we saw previously, like TFIID, TFIIIB is a protein complex consisting of TBP and TAFs. One of these TAFs, BRF, exhibits homology to TFIIB in both its C- and N-terminal domains. Certain mutations in the BRF N-terminal domain are known to greatly reduce the efficiency of class III promoter opening, while having no effect on PIC formation. Thus, the class III homolog of TFIIB may play a direct role in promoter opening. A similar role for TFIIB in class II promoter opening has not been uncovered: this could relate to the presence of an additional factor in the class II machinery (TFIIH) that appears to use the energy of ATP hydrolysis to drive promoter opening (see below).

3.6.3 Functional similarity between TFIIB family proteins and σ

In many ways, TFIIB family proteins are functionally analogous to bacterial σ factors. First, both $TFIIB_C$ and σ make important sequence-specific contacts with promoter DNA. Second, like domain 2 of σ, $TFIIB_N$ may play a role in recognizing the DNA close to the transcriptional start site and initiating promoter melting. Finally, like the $σ_3$–$σ_4$ linker, a part of $TFIIB_N$ may initially inhibit formation of the TEC by standing in the path of the growing nascent transcript.

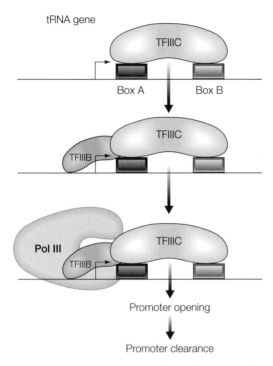

Figure 3.4 *Initiation at class III promoters.* The only essential promoter elements in most class III promoters (including the promoters of tRNA genes) are downstream of the start site, within the coding region. The promoter includes two conserved sequence blocks termed box A and box B. These form a binding site for TFIIIC, which then recruits TFIIIB to the region flanking the transcriptional start site. TFIIIB recruits Pol III yielding the complete preinitiation complex.

This is not meant to imply that TFIIB and σ factors have a common evolutionary ancestor – an idea that does not seem to be supported by sequence analysis. Rather the similarity in function between TFIIB and σ may represent a remarkable example of convergent evolution, in which two unrelated proteins have independently evolved to function in very similar ways.

3.7 TFIIH IN PROMOTER OPENING AND PROMOTER CLEARANCE

3.7.1 A unique DNA helicase requirement for class II promoter opening

DNA helicases couple ATP hydrolysis to the conversion of double-stranded DNA to single-stranded DNA. Pol II is the only major RNA polymerase

> **DNA helicases** – enzymes that sep-
> arate the strands in double-stranded
> DNA while hydrolyzing ATP. ATP
> hydrolysis may be coupled to changes
> in enzyme structure that drive the
> DNA strands apart.

> **What difference between Pol II
> and other RNA polymerases
> leads to the need for a DNA heli-
> case and ATP hydrolysis in class
> II promoter opening?**
>
> This requirement may serve to
> increase regulatory diversity but the
> structural features of the class II PIC
> that result in the need for a helicase
> are not understood.

that is unable to open the promoter without assistance from such enzymes. As a
result, formation of the Pol II open complex requires hydrolysis of ATP to ADP.
This requirement is unique to Pol II – it
is not observed for bacterial polymerases,
archaeal polymerases, Pol I, or Pol III.

The helicase activity required for class II
promoter opening is provided by TFIIH,
which is the last factor to join the PIC in
the classical assembly pathway. TFIIH is
a > 500 kD protein complex containing
10 distinct gene products (see Table 3.1).
The complete complex (sometimes termed
holo-TFIIH) consists of two dissociable
subcomplexes, termed core TFIIH and
cyclin-activating kinase (CAK). Core TFIIH
contains two DNA helicases, XPB and
XPD, and the catalytic activity of XPB is
essential for open complex formation.

As will be discussed below, TFIIH has
functions besides its helicase function. However, the helicase function is
the only one that is essential for transcriptional initiation. As a result,
TFIIH becomes dispensable for initiation when the energetic barrier to
promoter opening is relieved. For example, the use of negatively super-
coiled templates eliminates the requirement for TFIIH. This is because
negative supercoiling destabilizes DNA duplexes, thereby facilitating the
formation of single-stranded bubbles in the absence of a helicase.

3.7.2 A connection between DNA repair and transcription

In addition to its role in transcription, core TFIIH has a separate role in
nucleotide excision repair (NER). In this process, bulky covalent lesions
in the DNA, such as the pyrimidine dimers generated by UV irradiation,
are removed by the serial action of endonucleases, which introduce
single-stranded breaks into the DNA on either side of the lesion; DNA
helicases, which release the single-stranded DNA fragment containing the
lesion; DNA polymerase, which fills in the resulting gap; and DNA ligase,
which seals the remaining single-stranded break. XPB and XPD, both of
which are found in core TFIIH, provide the required helicase activities
for NER since protein extracts made from cells mutant in either of these
helicases are unable to mediate NER. The activity of these mutant extracts
can be restored by the addition of core TFIIH.

The discovery of a single protein complex required for both transcription and DNA repair provides a possible explanation for a phenomenon

known as transcription-coupled DNA repair, in which actively transcribed regions of the genome are repaired much more efficiently than transcriptionally silent regions. Apparently, the recruitment of TFIIH to a gene during initiation helps to nucleate the formation of repair complexes.

Mutations in the genes encoding the two helicase subunits in TFIIH, XPB and XPD, result in xeroderma pigmentosum, a heritable human disease associated with extreme sensitivity to sunlight and high susceptibility to skin cancer. Mutations in XPB and XPD can also result in Cockayne syndrome, which is associated with multiple developmental abnormalities such as mental retardation and skeletal abnormalities. While the light sensitivity in XP patients likely results from defective repair of UV damage to DNA, the developmental abnormalities in Cockayne syndrome may result from failures in transcription.

3.7.3 TFIIH and RNA polymerase II phosphorylation

As mentioned above, TFIIH consists of core TFIIH plus the CAK subcomplex. CAK contains a protein kinase of the cyclin-dependent kinase family termed Cdk7. As their name implies, cyclin-dependent kinases are only active when bound to proteins termed cyclins. The cyclin partner of Cdk7 (termed cyclin H), is also a component of CAK.

Transcriptional initiation is associated with the phosphorylation of the Pol II C-terminal domain (CTD), and it is the role of Cdk7 to catalyze this phosphorylation reaction. As mentioned in the previous chapter, the CTD is a feature of Pol II that is absent from all other known RNA polymerases. It is found at the C-terminal end of the largest subunit (the β' homologous subunit) and consists of multiple degenerate repeats of a heptapeptide sequence. While not essential for PIC assembly or RNA synthesis, the CTD is essential for viability. The CTD is highly mobile and as a result is not visible in any of the high resolution structures that have been determined for Pol II.

The CTD can be extensively phosphorylated, particularly on Ser2 and Ser5 of the heptapeptide (Figure 3.5), and the elongating polymerase contains an average of one phosphate group per heptapeptide repeat. However, the pool of free Pol II available for PIC assembly is apparently in the unphosphorylated state. This is probably important for the proper regulation of gene expression, since a protein complex called the Mediator, which plays critical roles in gene regulation (Chapter 4), preferentially binds the unphosphorylated form of Pol II.

Transition from the open complex to the TEC is accompanied by the extensive phosphorylation of the CTD, mostly on Ser5, by Cdk7. While this phosphorylation reaction is not required for transcriptional initiation, it may convert the CTD into a docking site for factors required for RNA processing. For example, phosphorylation of the CTD at Ser5 increases the affinity of the CTD for the capping enzyme, which is responsible for

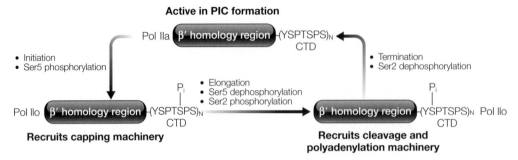

Figure 3.5 *The RNA polymerase II phosphorylation cycle.* Most of the free Pol II in the nucleus is in the unphosphorylated form (top). Unphosphorylated Pol II binds to the promoter along with the general transcription factors leading to initiation. Very shortly after initiation (probably coinicident with promoter clearance), Cdk7 phosphorylates the Pol II CTD on Ser5. The Ser5 phosphorylated form of the CTD (bottom, left) serves as a platform for the assembly of the 5' capping machinery. As elongation proceeds, Ser5 is dephosphorylated, while Ser2 becomes phosphorylated. The Ser2 phosphorylated form of the CTD (bottom, right) serves as a platform for the recruitment of the 3' cleavage and polyadenylation machinery. After termination, Ser2 is rapidly dephosphorylated. The unphosphorylated form of Pol II is termed Pol IIa, while the various phosphorylated forms of Pol II are termed Pol IIo.

adding a special structure (the 7-methyl-Gppp cap) to the 5' end of all class II transcripts. This cap serves both to protect class II transcripts from 5' to 3' exonucleolytic degradation and to target the transcript to the ribosome. Phosphorylation at Ser5 at initiation leads to the efficient recruitment of capping enzyme, which then catalyzes efficient capping of the growing transcript.

As transcription proceeds, the pattern of CTD phosphorylation changes (Figure 3.5). In particular, CTD phosphatases remove the phosphate groups from Ser5, while Ser2-specific kinases add phosphate groups to the CTD. As a result, while the CTD in the elongating polymerase is primarily phosphorylated at Ser5 in promoter proximal regions of transcription units, it is primarily phosphorylated at Ser2 in promoter distal regions. Just as the Ser5 phosphorylated CTD serves as a platform for the assembly of the 5' capping machinery, the Ser2 phosphorylated CTD serves as a platform for the assembly of the machinery that catalyzes cleavage and polyadenylation at the 3' end of the transcript.

Thus, the coordinated phosphorylation of the CTD first on Ser5 by the TFIIH subunit Cdk7 and later on Ser2 by other kinases helps to coordinate the overall process of mRNA biogenesis, ensuring that only Pol II transcripts will be subject to the post-transcriptional modifications (5' capping, and 3' cleavage and polyadenylation) that mark protein-encoding transcripts.

3.8 SUMMARY

While each eukaryotic and archaeal RNA polymerase contains its own set of general transcription factors, the sets overlap strongly hinting at conserved mechanisms for promoter recognition, start site selection, and promoter melting. TBP is a universal transcription factor that binds TATA boxes and is required for transcription by all eukaryotic and archaeal promoters regardless of whether or not they contain a TATA box. Eukaryotic cells possess three TBP-containing complexes, one for the transcription of each class of genes. The class II complex TFIID contains ~14 TAFs in addition to TBP, and forms a trilobular horseshoe-shaped structure that may straddle the DNA. These TAFs have multiple functions including functions in class II promoter recognition and transcriptional activation.

While TFIID is primarily responsible for recognizing the promoter, TFIIB serves as a bridge between TBP and RNA polymerase II. Both the class III machinery and the archaeal machinery contain TFIIB homologs (BRF and TFB, respectively). In addition to serving as a bridge between TBP and RNA polymerase, TFIIB family proteins appear to play roles in promoter recognition, start site selection, and DNA melting. In many respects, the functions of TFIIB are reminiscent of the functions of bacterial σ factors, although this probably represents an example of convergent evolution since there is little if any sequence homology between TFIIB and σ factors.

Unlike all other transcription systems, the Pol II system requires a high energy cofactor, namely ATP, for open complex formation. A helicase subunit in TFIIH termed XPB couples hydrolysis of ATP to promoter opening at class II promoters. TFIIH is also a component of the nucleotide excision repair machinery and may be responsible for transcription coupled DNA repair.

TFIIH also contains a module termed CAK that includes the Cdk7 kinase. Cdk7 phosphorylates the Pol II CTD on Ser5 during promoter clearance. As elongation proceeds, Ser5 is dephosphorylated and other kinases phosphorylate Ser2. The Ser5 phosphorylated CTD at the promoter proximal end of the transcription unit recruits the 5' end mRNA processing machinery, while the Ser2 phosphorylated CTD at the promoter distal end of the transcription unit recruits the 3' end mRNA processing machinery. Thus, CTD phosphorylation and dephosphorylation help to coordinate the steps in mRNA biogenesis.

PROBLEMS

1 Unlike all other known polymerases, Pol II appears to require a helicase (XPB) for promoter opening. How might the exceptional nature of Pol II in this regard be related to the exceptionally high regulatory demands on the class II general machinery?

2 It is possible that TBP binds directly to TATA-less promoters (albeit in a non-specific manner) and that this binding is required for PIC formation at TATA-less promoters. Alternatively, it is possible that direct binding of TBP to TATA-less promoters is not required for PIC formation at TATA-less promoters.

(a) Design an experiment to distinguish between the two possibilities presented above.

(b) Assuming direct binding of TBP to DNA is not required for transcription from TATA-less promoters, how do you explain the TBP requirement for transcription from these promoters?

3 Although TFIIB is functionally similar to σ, it bears no sequence homology to σ. In contrast, TFIIF has been reported to exhibit limited sequence homology to σ. In what way is TFIIF functionally analogous to σ?

FURTHER READING

The class II machinery and preinitiation complex assembly

Orphanides, G., Lagrange, T. and Reinberg, D. (1996) The general transcription factors of RNA polymerase II. *Genes Dev*, **10**, 2657–2683. *A comprehensive review of the class II general transcriptional machinery.*

Hahn, S. (2004) Structure and mechanism of the RNA polymerase II transcription machinery. *Nat Struct Mol Biol*, **11**, 394–403. *A recent review on the class II machinery emphasizing structure/function relationships.*

Juven-Gershon, T., Hsu, J.Y. and Kadonaga, J.T. (2006) Perspectives on the RNA polymerase II core promoter. *Biochem Soc Trans*, **34**, 1047–1050. *A succinct review of the class II core promoter.*

Matsui, T., Segall, J., Weil, P.A. and Roeder, R.G. (1980) Multiple factors required for accurate initiation of transcription by purified RNA polymerase II. *J Biol Chem*, **255**, 11992–11996. *The earliest demonstration that eukaryotic cells contain multiple distinct general factors required for initiation from class II promoters.*

Buratowski, S., Hahn, S., Guarente, L. and Sharp, P.A. (1989) Five intermediate complexes in transcription initiation by RNA polymerase II. *Cell*, **56**, 549–561. *A classic study working out the order of assembly of the components of the class II preinitiation complex.*

TFIID

Kim, J.L., Nikolov, D.B. and Burley, S.K. (1993) Co-crystal structure of TBP recognizing the minor groove of a TATA element. *Nature*, **365**, 520–527. *One of the first published structures of the TBP/TATA box complex showing the unusual distortion of the target DNA.*

Andel, F., 3rd, Ladurner, A.G., Inouye, C., Tjian, R. and Nogales, E. (1999) Three-dimensional structure of the human TFIID-IIA-IIB complex. *Science*, **286**, 2153–2156. *The TFIID "horseshoe" structure.*

Leurent, C., Sanders, S., Ruhlmann, C., Mallouh, V., Weil, P.A., Kirschner, D.B., Tora, L. and Schultz, P. (2002) Mapping histone fold TAFs within yeast TFIID. *EMBO J*, **21**, 3424–3433. *Mapping of TAFs in TFIID by immunoelectron microscopy.*

Chen, J.L., Attardi, L.D., Verrijzer, C.P., Yokomori, K. and Tjian, R. (1994) Assembly of recombinant TFIID reveals differential coactivator requirements for distinct transcriptional activators. *Cell*, **79**, 93–105. *The role of TAFs as targets of transcriptional activators.*

TFIIB

Littlefield, O., Korkhin, Y. and Sigler, P.B. (1999) The structural basis for the oriented assembly of a TBP/TFB/promoter complex. *Proc Natl Acad Sci USA*, **96**, 13668–13673. *Demonstration of how TBP/TFIIB/promoter interactions can serve to determine the direction of transcription.*

Pinto, I., Wu, W.H., Na, J.G. and Hampsey, M. (1994) Characterization of sua7 mutations defines a domain of TFIIB involved in transcription start site selection in yeast. *J Biol Chem*, **269**, 30569–30573. *Use of genetic suppression to demonstrate a role for TFIIB in start site selection.*

Chen, H.T. and Hahn, S. (2004) Mapping the location of TFIIB within the RNA polymerase II transcription preinitiation complex: a model for the structure of the PIC. *Cell*, **119**, 169–180. *Elegant study mapping the sites of interaction between TFIIB and RNA polymerase II.*

Hahn, S. and Roberts, S. (2000) The zinc ribbon domains of the general transcription factors TFIIB and Brf: conserved functional surfaces but different roles in transcription initiation. *Genes Dev*, **14**, 719–730. *Demonstration that a TFIIB paralog in the class III basal machinery is required for promoter opening.*

TFIIH and CTD phosphorylation

Tirode, F., Busso, D., Coin, F. and Egly, J.M. (1999) Reconstitution of the transcription factor TFIIH: assignment of functions for the three enzymatic subunits, XPB, XPD, and cdk7. *Mol Cell*, **3**, 87–95. *Biochemical demonstration of the roles of TFIIH in promoter opening and CTD phosphorylation.*

Komarnitsky, P., Cho, E.J. and Buratowski, S. (2000) Different phosphorylated forms of RNA polymerase II and associated mRNA processing factors during transcription. *Genes Dev*, **14**, 2452–2460. *Demonstration of the dynamic nature of the phosphorylation state of the Pol II CTD during the transcription cycle and of the role of phosphorylated CTD in the recruitment of capping enzyme.*

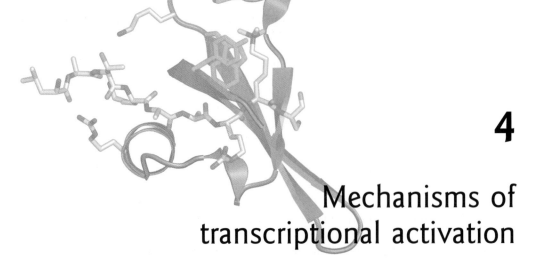

4

Mechanisms of transcriptional activation

Key concepts

- Activators recognize cis-elements via sequence-specific α-helix major groove contacts
- Activators signal through many target surfaces in the transcriptional machinery to stimulate the transcription cycle
- Multiple steps in the transcription cycle, including RNA polymerase binding, promoter opening, and elongation, are subject to regulation by activators

4.1 INTRODUCTION

The general transcriptional machinery usually requires help from activator proteins to recognize and use promoters in an efficient manner. The inability of core promoters to direct transcription on their own is not an indication of evolutionary imperfection, but instead provides an opportunity for gene regulation, which occurs when regulatory factors stimulate transcription in response to developmental or environmental cues.

This chapter will begin with a discussion of two bacterial transcription factors, the bacteriophage λcI protein and the *Escherichia coli* cyclic AMP (cAMP) receptor protein, which have served as paradigms in our efforts to understand mechanisms of transcriptional activation. This will be followed by an introduction to eukaryotic activator proteins. A major message of this chapter is that the mechanisms by which activators communicate with the basal machinery are conserved across all domains of life. This communication requires contacts between activators and the transcriptional machinery to bring RNA polymerase to the promoter and stimulate subsequent steps in the transcription cycle.

While bacterial and eukaryotic transcription factors use conserved strategies to signal to the basal machinery, there are also significant differences in the mechanisms of bacterial and eukaryotic activation. First, eukaryotic cells contain protein complexes (coactivators, Chapter 1) not found in bacteria that function to relay signals from activators to the general transcriptional machinery. One such complex is termed the Mediator. Second, as we saw in the previous two chapters, eukaryotic RNA polymerase II contains a unique feature termed the C-terminal domain (CTD), which is phosphorylated and dephosphorylated during each transcription cycle. The presence of this protein kinase target in the transcriptional machinery provides regulatory opportunities that do not exist in bacteria. Third, eukaryotic activators are able to communicate with enzymes that modulate chromatin structure, thereby modulating the efficiency with which genes are recognized by the basal machinery. While the Mediator and CTD phosphorylation will be discussed in this chapter, the role of chromatin structure in gene regulation will be the subject of Chapters 5 and 6.

4.2 PARADIGMS FROM *E. COLI*

4.2.1 CRP: a sensor of the nutritional environment

One of the best studied of all activators, and one upon which much of our knowledge about mechanisms of activation is based, is the *E. coli* cAMP receptor protein (CRP), which is also known as the catabolite gene activator protein (CAP). This factor activates operons required for the conversion of a variety of sugars (lactose, galactose, maltose, etc.) to glucose.

Cyclic AMP (cAMP) – an important second messenger in both bacteria and eukaryotes formed from ATP in response to a variety of extracellular signals by the enzyme adenylate cyclase. In bacteria, cAMP is an allosteric activator of CRP, while in eukaryotes, it activates protein kinase A, which then phosphorylates and activates a transcription factor termed cAMP response element-binding protein (CREB).

CRP binds cAMP, and only activates transcription when bound to this nucleotide. Since the enzyme that synthesizes cAMP (adenylate cyclase) is inactivated by glucose, CRP is only active in the absence of glucose (Figure 4.1A). In the presence of glucose, CRP is inactive and operons required for the metabolism of sugars other than glucose are not transcribed. This is part of a strategy to conserve energy since there is no need to synthesize the gene products that convert sugars such as lactose into glucose if glucose is already in abundant supply.

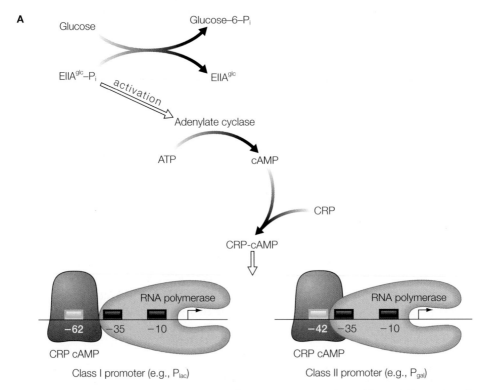

Figure 4.1 *cAMP receptor protein (CRP) and λcI-dependent genetic switches.* (A) CRP regulates expression of the lac and gal operons in response to glucose availability. CRP activity requires cAMP, which is synthesized by adenylate cyclase. This enzyme is activated by the phosphorylated form of the EIIA^{glc} protein. cAMP binds CRP allowing it to bind and turn on class I CRP responsive promoters such as P$_{lac}$ and class II CRP responsive promoters such as P$_{gal}$. But in the presence of glucose, EIIBC^{glc} transfers the phosphate group of EIIA^{glc} to glucose; adenylate cyclase is therefore inactive. As a result, the P$_{lac}$ and P$_{gal}$ promoters remain off.

4.2.2 λcI: regulator of the lysis/lysogeny switch

Many sequence-specific transcription factors can function as either activators or repressors depending on binding site context or the cellular environment. The cI protein of bacteriophage λ (a double-stranded DNA bacteriophage that infects *E. coli*), was one of the first factors known to play both roles. While λcI will be discussed extensively in this chapter, a number of other factors capable of functioning as both activators and repressors (e.g., nuclear receptors) are discussed in Chapter 7.

Whenever λ infects an *E. coli* cell, it can enter either the lytic or lysogenic growth mode. In lytic growth, the phage replicates and induces lysis of the host cell releasing progeny phage into the environment, while in lysogenic growth, the phage genome integrates into the host chromosome and is replicated along with the host DNA. Lysogenic growth requires the λcI protein, which binds to several operators and represses transcription of genes required for lytic growth. The λcI protein also activates the promoter of its own gene, thereby stabilizing the lysogenic growth mode. A variety of environmental insults trigger the destruction of the λcI protein leading to lytic growth.

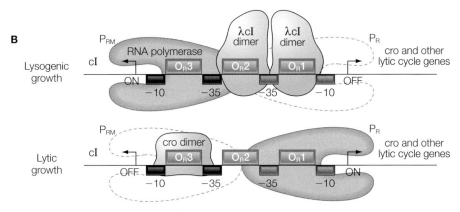

Figure 4.1 (*continued*) *cAMP receptor protein (CRP) and λcI-dependent genetic switches.* (B) λcI controls the switch between the λ bacteriophage lytic and lysogenic states. These pictures illustrate the control region between the two λ promoters P_R and P_{RM}, which contains three operators termed O_R1, O_R2, and O_R3. Lytic growth requires transcription from P_R, while lysogenic growth requires that P_R be switched off. Transcription of P_{RM} gives rise to cI protein, which shuts off P_R and also activates its own expression. During lysogenic growth (top), the cI protein cooperatively binds to O_R1 and O_R2. cI bound to O_R1 sterically excludes RNA polymerase from P_R, while cI bound to O_R2 contacts RNA polymerase bound to P_RM, activating this promoter. Destruction of cI leads to depression of P_R and entry into the lytic growth cycle (bottom). During lytic growth, cro protein (which is one of the lytic genes expressed from P_R) binds to O_R3 and represses P_{RM}.

The lysis/lysogeny switch is controlled by the activity states of two promoters termed P_R and P_{RM}, which are reciprocally regulated by λcI. P_R gives rise to a transcript that encodes proteins required for lytic growth, while P_{RM} gives rise to a transcript encoding the λcI protein. P_R and P_{RM} are only about 80 bp apart and point away from one another (Figure 4.1B). Between them are three λcI-binding sites termed O_R1, O_R2, and O_R3. During lysogenic growth, λcI protein occupies both O_R1 and O_R2. λcI bound to O_R1 excludes RNA polymerase from P_R, preventing the switch to lytic growth. At the same time, λcI bound to O_R2 activates P_{RM}, resulting in the continued synthesis of λcI and therefore stabilizing the lysogen. The mechanism by which λcI activates P_{RM} will be explored in detail later in this chapter.

4.2.3 Sequence-specific DNA recognition: the helix-turn-helix motif

Activators are directed to specific promoters via sequence-specific DNA recognition and binding. Unlike proteins, the three-dimensional structures of which are extremely sensitive to amino acid sequence, the path of the sugar–phosphate backbone in double-stranded DNA varies little from one DNA sequence to the next. Therefore, to discriminate between DNA sequences, DNA-binding proteins must read the base sequence directly by contacting the edges of the bases through either the major or minor groove of the DNA (Plate 4.1A). These contacts most often take the form of hydrogen-bond interactions. Sequence-specific binding occurs when a surface on the protein contains a spatial array of H-bond donors and acceptors that is complementary to the spatial array of H-bond donors and acceptors present in one of the grooves of the DNA.

AT and GC basepairs present distinguishable arrays of H-bond donors and acceptors in both the major and minor grooves (Plate 4.1B). Thus, sequence-specific DNA-binding proteins can recognize different sequences by binding in either groove of the DNA. However, the major groove is a much richer source of sequence information, because the sequence-specific differences in the geometric arrangements of the H-bond donors and acceptors are much more dramatic in the major than in the minor groove.

Both CRP and λcI utilize a protein motif termed the helix-turn-helix (HTH) motif to recognize and bind specific regulatory elements in the DNA. HTH motifs are found in most bacterial transcription factors, as well as in a significant fraction of eukaryotic transcription factors. As the name implies, HTH motifs consist of two α-helices separated by a turn (Plate 4.2A, B). The two helices pack against one another at roughly right angles, and the second helix (the recognition helix) sits in the major groove of the DNA. The α-helix and the major groove are complementary in shape and thus the insertion of an α-helix into the major groove results in extensive contact between the protein and DNA surfaces. Different recognition helices recognize and bind different DNA sequences via specific H-bond contacts between the amino acid side chains in the recognition helix and the atoms on the major groove edges of the base pairs (Plate 4.2B, C).

Precise positioning of the recognition helix in the major groove is the job of the first helix in the HTH motif. This helix packs against the recognition helix and also makes electrostatic and H-bond contacts with the sugar–phosphate backbone of the DNA, stabilizing the recognition helix in the correct position and orientation (Plate 4.2B, C). Because of its role in stabilizing the recognition helix in the major groove, the first helix of the HTH motif has been termed a "molecular outrigger" by analogy with the outrigger on a canoe.

4.2.4 Dimerization

Like many other bacterial activators, both λcI and CRP must dimerize before they can bind DNA with significant affinity and specificity (Plates 4.2A, 4.3). Dimerization doubles the protein surface area in contact with the DNA, thereby doubling the free energy released upon formation of the protein–DNA complex. Given the exponential relationship between equilibrium constants and standard free energy doubling the free energy squares the equilibrium constant of the complex. The association of two λcI or CRP monomers to form a symmetric homodimer results in the formation of a symmetric DNA interaction surface. As a result, these factors both bind to symmetric recognition elements consisting of two half-sites in inverted repeat orientation (Plate 4.3).

Each CRP subunit consists of two globular domains connected by an apparently flexible hinge region (Plate 4.3). The N-terminal domain mediates dimerization, while the C-terminal domain contains the HTH motif that contacts DNA. The CRP N-terminal domain also binds cAMP with one molecule of cAMP binding to each subunit. It is not understood how the binding of cAMP alters the structure of CRP to allow DNA binding.

> **Half-sites** – Homodimeric transcription factors such as CRP and λcI usually bind to DNA elements consisting of two half-sites in inverted repeat orientation. Each subunit of the homodimer contacts one half-site in the recognition element.

4.2.5 Multiple targets for activators in the transcriptional machinery

Positive control mutations

To learn about the precise mechanisms by which λcI and CRP stimulate transcription, geneticists have generated altered forms of these factors that are unable to activate transcription, and then carefully analyzed the exact nature of the resulting transcriptional defects. In the case of λcI, a genetic screen was devised to identify mutant forms of the factor that could still repress transcription from P_R, but that could not activate transcription from P_{RM}. Such mutations were termed positive control (pc) mutations. Since the pc mutants could still repress transcription it was reasoned that their DNA-binding function must still be intact, and that the mutations must prevent the bound activator from signaling to RNA polymerase. Remarkably, sequencing of the mutations revealed that all the pc mutations mapped close to one another in the HTH motif (Plate 4.2B), suggesting that the same motif responsible for sequence-specific contacts with DNA is also directly responsible for communication with RNA polymerase.

Mutations in CRP analogous to the pc mutations in λcI have also been identified. The CRP pc mutations define at least two separate surfaces termed activation region 1 (AR1) and activation region 2 (AR2) (Plate 4.3). For some CRP-responsive promoters (e.g., the lac promoter) AR1 is sufficient for activation, while for other CRP-responsive promoters (e.g., the gal promoter), both AR1 and AR2 are required for activation.

Activator–RNA polymerase interactions

Since the pc mutations define surfaces required for communication with polymerase, it is reasonable to ask if these surfaces are in direct contact with the RNA polymerase holoenzyme. This question has been addressed using both genetic (Box 4.1) and biochemical (Box 4.2) approaches, and the results show that each activation surface contacts a distinct surface in RNA polymerase. In particular, the activation surface defined by the λcI pc mutations contacts the σ subunit, while AR1 and AR2 in CRP each contact distinct surfaces in the α subunit.

An important message of these studies of activation is that activators have evolved to interact with multiple surfaces in the transcriptional machinery. Studies of just two activators reveal three different interaction surfaces in the RNA polymerase holoenzyme (one in σ and two in α). This is almost certainly just the tip of the iceberg. The availability of multiple interaction surfaces within the transcriptional machinery allows for the possibility of multiple activators contacting the machinery simultaneously. This, in turn, can allow multiple activators to cooperate with one another in activating transcription. As will be seen in Chapter 7, cooperation between activators is probably much more important in eukaryotes where multiple activators usually work together to turn on a gene than in bacteria where each gene is often regulated by a single activator.

4.2.6 How do activator–RNA polymerase contacts activate promoters?

Measuring the effects of activators on the kinetics of initiation

The findings described above indicate that direct contacts between bacterial activators and RNA polymerase are required for activation. But how

Box 4.1

Protein–protein interactions and allele-specific genetic suppression

Studies to learn about transcriptional mechanisms often depend on being able to determine what touches what. One approach that can be employed to characterize such interactions is allele-specific genetic suppression. This begins with a strain containing a mutation in a gene of interest (gene A) that results in a readily detectable phenotype. The genome is then mutagenized and the mutants are screened to find mutations in other genes that eliminate (or suppress) the phenotype due to the original mutation. A mutation of this type (termed a second site suppressor mutation) defines a gene (gene B) that may interact with gene A. These interactions can be of many types and can be very indirect. However, a strong case for a direct interaction between the products of the two genes can be made if the suppression is allele specific. In allele-specific suppression, the second site suppressor mutation only suppresses the phenotype due to the mutant allele of gene A employed in the screen and not the phenotype due to other mutant alleles of gene A.

The idea of allele-specific suppression can be readily understood by thinking of the products of the interacting genes as geometric objects with complementary shapes that allow them to bind one another (Figure B4.1A). The original mutation in gene A results in a change in the shape of its product that prevents binding to the product of gene B, while the mutation in gene B changes the shape of its product in such a way as to restore binding to the product of gene A.

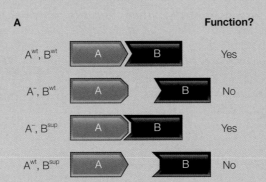

Figure B4.1 *Screen for an allele-specific suppressor.* (A) In allele-specific suppression, a suppressor allele of gene B (B^{sup}), allows function in the presence of particular mutant allele of gene A because the suppressor mutation restores the complementarity between the products of the two genes. B^{sup} does not function in the presence of other alleles of gene A such as the wild-type allele (A^{wt}).

A beautiful example of this type of allele-specific suppression comes from studies of the λcI protein (Figure B4.1B). These studies began with a positive control mutant of the λcI protein. The subunits of RNA polymerase were then mutagenized and a single mutant allele of the gene encoding the σ subunit was found that could suppress the positive control mutation. The suppression was allele specific, strongly suggesting a direct contact between λcI and σ.

B Activator protein	Activation of P_{RM}	
	σ^{wt}	σ^{sup}
λcI wild-type	+	−
λcI pc1	−	−
λcI pc2	−	+
λcI pc3	−	−

Figure B4.1 (*continued*) *Screen for an allele-specific suppressor.* (B) Mutagenesis of the σ subunit of RNA polymerase revealed an allele (σ^{sup}) that allowed the pc2 allele of λcI to activate P_{RM}. However, other alleles of λcI failed to activate P_{RM} in the presence of σ^{sup}. This allele-specific suppression argues strongly for a direct contact between the activation surface of λcI and the σ subunit of RNA polymerase.

do these contacts stimulate transcription? As explained in Chapter 2, the transcription cycle can be divided into multiple steps, including an initiation phase, an elongation phase, and a termination phase. These phases can, in turn, be divided into multiple subphases. For example, initiation includes closed complex formation, open complex formation, abortive initiation, and promoter clearance, and activators could work at any of these steps. Therefore, an understanding of the mechanisms by which activators work requires methods for measuring the rate and equilibrium constants associated with each step.

An oversimplified, but still useful way to view open complex formation is summarized in the follow scheme:

$$R + P \underset{}{\overset{K_B}{\rightleftharpoons}} RP_C \overset{k_f}{\longrightarrow} RP_O$$

where R is the free RNA polymerase, P is the free promoter, RP_C is the closed promoter complex, and RP_O is the open promoter complex. K_B is the equilibrium constant for formation of the closed complex, while k_f is the rate constant for conversion of the closed complex to the open complex. One well-established technique for measuring K_B and k_f is the abortive initiation assay (Box 4.3). These measurements have shown that activators can alter both the stability of the closed complex (K_B) and the rate of open complex formation (k_f). For example, assays carried out to measure

Box 4.2

Using chemical probes to detect protein interactions

Box 4.1 describes a geneticist's approach to defining protein–protein interactions. An alternative approach favored by biochemists is the use of chemical probes, such as cross-linking reagents, to detect interactions in a test tube. A cross-linking reagent is attached to one protein, which is then mixed with other proteins with which the first protein is suspected to interact. If an interaction occurs, then the cross-linker will be brought into close proximity with the interacting protein. A reaction can then be induced in which the cross-linker becomes attached to the second protein thereby covalently linking the two proteins to one another. This is the approach that was used to detect the interaction between CRP and the α subunit of RNA polymerase. An overview of the approach used is shown in Figure B4.2A and the structure of the specific cross-linker employed as well as the chemistry involved in its use is shown in Figure B4.2B.

The cross-linking reagent used to demonstrate the CRP–RNA polymerase interaction was designed so that the linkage to CRP could be easily reversed. Thus, after the cross-link was formed, the link to CRP could be broken, with the net result being the transfer of the cross-linker from CRP to RNA polymerase. Since the cross-linker contained a radioactive iodine atom, it was possible to determine where within RNA polymerase the CRP contact occurred simply by determining which subunit within RNA polymerase had acquired the radioactivity. This variation on the basic cross-linking experiment is often termed a "label transfer" experiment.

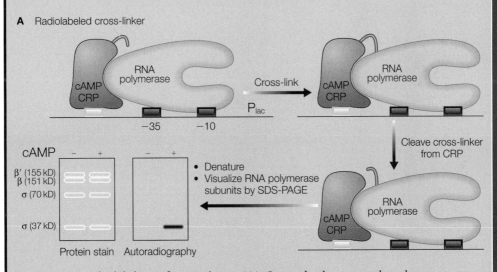

Figure B4.2 *The label transfer experiment.* (A) General scheme used to demonstrate an interaction between CRP and RNA polymerase. The illustration at bottom left is a schematic representation of the results of an SDS polyacrylamide gel electrophoresis (SDS-PAGE), which separates polypeptides on the basis of size. The panel on the left shows a gel stained with a dye that binds all proteins and we can therefore see bands representing all the subunits of RNA polymerase. The panel on the right shows an autoradiogram of the same gel, in which only radioactive polypeptides are detected.

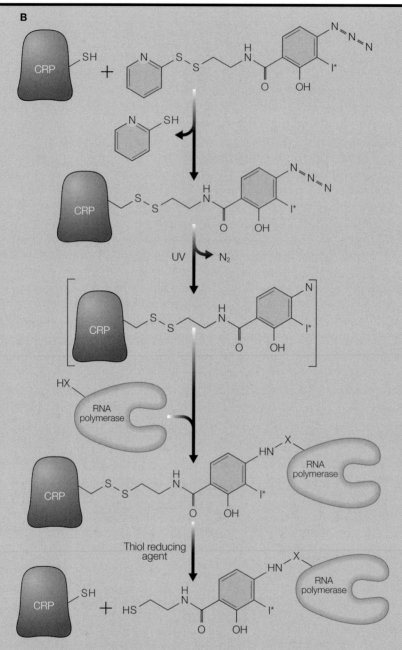

Figure B4.2 (*continued*) *The label transfer experiment.* (B) A detailed look at the chemistry used to achieve cross-linking and then label transfer. The cross-linking reagent is attached to CRP via a disulfide bond. Irradiation with UV results in the elimination of N_2. The transient intermediate (shown in square brackets) is highly reactive because the nitrogen atom is only making a single covalent bond, whereas a stable neutral nitrogen atom prefers to make three covalent bonds. It therefore reacts rapidly with any nearby atoms. If another protein is nearby it can become covalently attached to this other protein thereby generating a cross-link. The disulfide bond to CRP can be subsequently broken by treatment with a reducing agent.

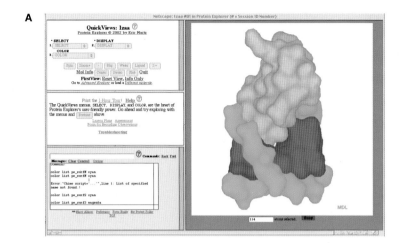

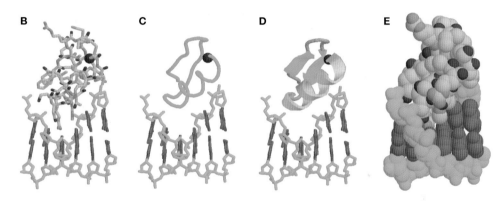

Plate B1.1 *Illustrating protein structure with Protein Explorer.* (A) The Protein Explorer window. The frame on the right contains the structure being manipulated, in this case a single zinc finger from the transcription factor Zif268 bound to DNA (PDB id 1ZAA). The protein and DNA are displayed in the surface view. The top left frame is the "quick views" frame, which contains several pull down menus that allow you to select, render, and color the molecules in many ways. (B–E) The Zif268 zinc finger rendered in various ways using Protein Explorer. (B–D) The DNA is shown in the wireframe view, while the protein is shown in either the wireframe view (B), the backbone trace view (C), or the cartoon view (D). (E) The DNA and protein are shown in the spacefilling view.

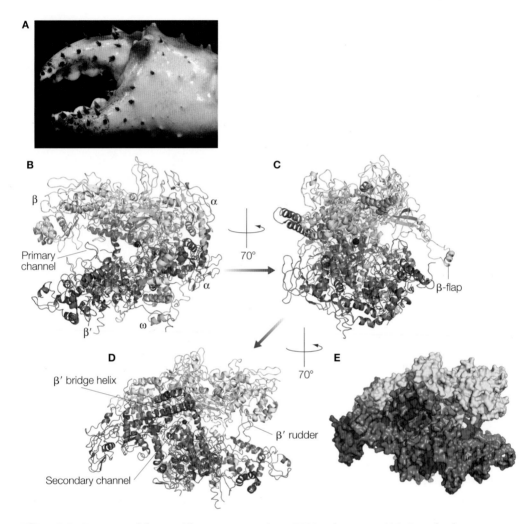

Plate 2.1 *Structure of the core* Thermus aquaticus *RNA polymerase.* (A) A crab claw.
(B–D) Backbone traces of the core polymerase, with various features (the primary and
secondary channels, the β-flap, the β′ rudder, and the β′ bridge helix) labeled (PDB id
1HQM). Relative rotation of B, C, and D with respect to one another is indicated. Note
that from the point of view shown in B, the polymerase resembles a crab claw. The
catalytic metal ion (metal A), which marks the active site, is shown as a red sphere.
(E) A surface representation from the same point of view as the backbone trace in D.
Note that although the active site metal ion is deep inside the polymerase, it is still
visible in the surface view through the secondary channel.

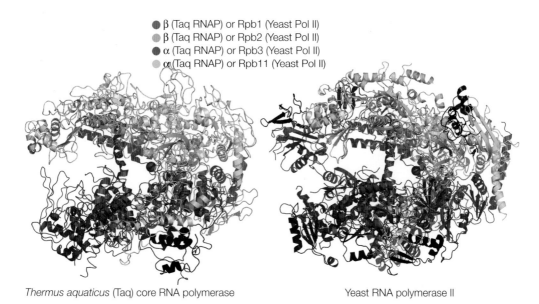

β (Taq RNAP) or Rpb1 (Yeast Pol II)
β (Taq RNAP) or Rpb2 (Yeast Pol II)
α (Taq RNAP) or Rpb3 (Yeast Pol II)
α (Taq RNAP) or Rpb11 (Yeast Pol II)

Thermus aquaticus (Taq) core RNA polymerase   Yeast RNA polymerase II

Plate 2.2 *Similar structures of bacterial and eukaryotic RNA polymerases.* The *T. aquaticus* (Taq) core polymerase is shown on the left (PDB id# 1HQM), while yeast RNA polymerase II is shown on the right (PDB id# 1I6H). The α, β, and β′ subunits of the bacterial enzyme and the homologous subunits in the yeast enzyme are color coded as indicated in the key at the top. The remaining subunits are colored black.

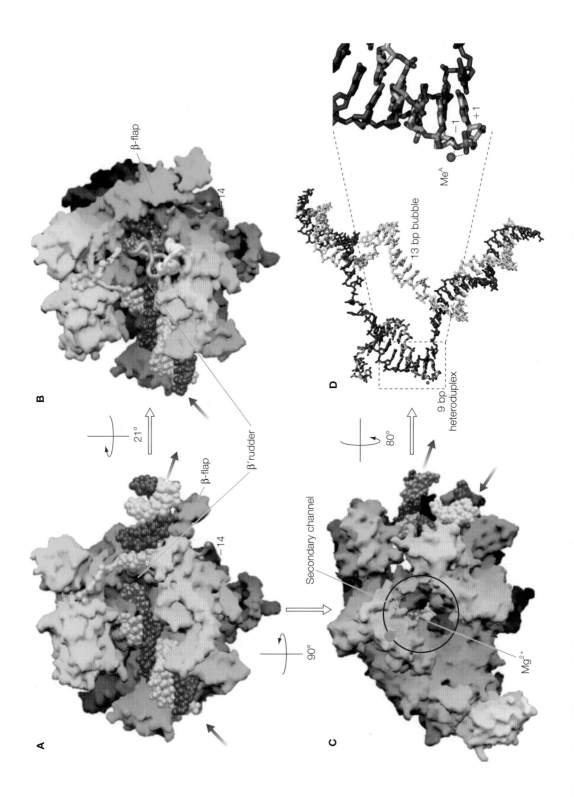

β-flap

−14

β-flap

β'rudder

21°

A

90°

Secondary channel

Mg²⁺

C

B

D

80°

9 bp.
heteroduplex

13 bp bubble

Me^A

−1

+1

Plate 2.3 (*opposite page*) *Model of the* T. aquaticus *ternary elongation complex.* This model is based on the X-ray crystal structure of the core polymerase into which the DNA and RNA have been docked. The position of the nucleic acids relative to the protein were determined by an extensive series of cross-linking studies to map contacts between nucleotide and amino acid residues. (A, B) Two slightly different frontal views of the polymerase looking down the primary channel between the two crab claws. The protein is shown in the surface view (β, cyan; β′, pink; α and ω, white). In (B) the β′ rudder is shown as a backbone trace. Metal A is visible as a sphere at the back of the primary channel in (B). The DNA template strand is red, the non-template strand is yellow, and the RNA is orange. Double-stranded nucleic acids are shown in the spacefilling view, while single-stranded nucleic acids are shown as backbone traces. The arrows indicate the direction that the template moves relative to the protein during elongation. Part of the DNA upstream of the transcription bubble has been removed in (B) for clarity. (C) A side view of the polymerase illustrating the access to the active site through the secondary channel. (D) A view in which the protein has been stripped away leaving just the DNA, RNA, and metal A. The nucleic acids are shown in wireframe view. The coloring of the DNA is the same as in (A–C), but the RNA has been rendered according to the CPK coloring scheme (see Box 1.2). In the magnified view of the heteroduplex (right), the position of the metal ion close to the final phosphodiester bond in the RNA is apparent. Thus, this model represents a view of the polymerase after phosphodiester bond formation but before translocation. A–C, from Korzheva, N. et al. (2000) *Science*, **289**, 619–625, with permission from AAAS; D, drawn using coordinates provided by Seth Darst (The Rockefeller University).

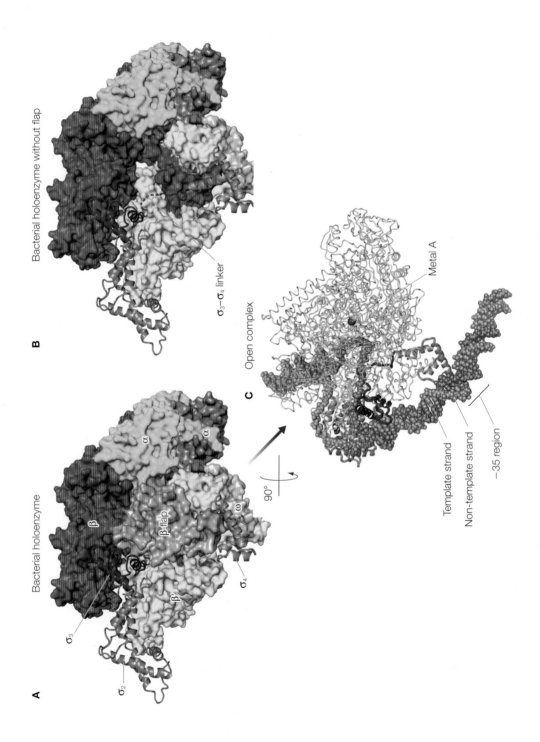

A Bacterial holoenzyme

β
σ₃
β'
β-flap
α
α
ω
σ₄
σ₂

B Bacterial holoenzyme without flap

σ₃–σ₄ linker

90°

C Open complex

Metal A

Template strand
Non-template strand
–35 region

Plate 2.4 (*opposite page*) *The structure of the* T. aquaticus *holoenzyme and open complex.* (A, B) X-ray crystal structure of the T. aquaticus RNA polymerase σ^A holoenzyme. σ^A is the T. aquaticus homolog of σ^{70}. The viewpoint is roughly equivalent to that shown for the core polymerase in Plate 2.1B. The core subunits are shown in surface view, while σ^A is shown in cartoon view. The σ_2, σ_3, and σ_4 domains are indicated. The σ_1 domain is blurred from view due to its mobility. In (B), the β-flap has been eliminated from view to reveal the σ_3–σ_4 linker, which runs underneath the flap. Part of this linker is invisible due to its mobility and therefore a dashed line connecting σ_3 and σ_4 has been drawn in to indicate the continuity of the protein between these two domains. (C) A model of the open complex based on the crystal structure of the holoenzyme, the crystal structure of the closed complex, and the model of the TEC. The DNA is shown in spacefilling view, while the protein is shown as a backbone trace, with σ^A colored as in A and B, and the core polymerase colored gray. A–C, drawn using coordinates provided by Seth Darst (The Rockefeller University).

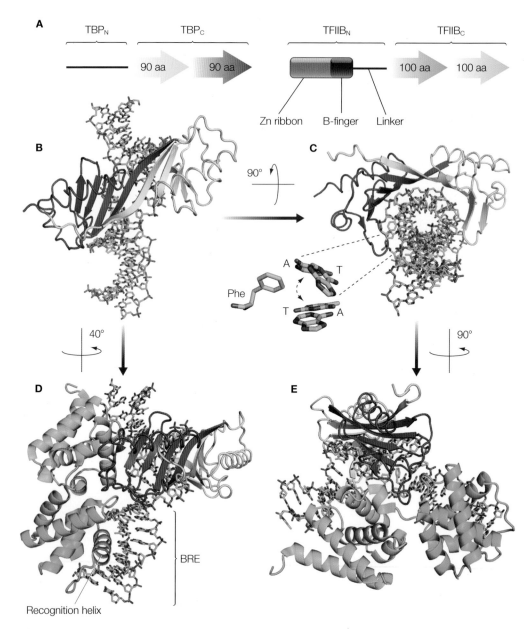

Plate 3.1 *The TBP$_C$/TFIIB$_C$/DNA complex* (PDB id 1D3U). (A) Schematic representations of the TATA-binding protein (TBP) and TFIIB primary structures. Each factor contains a conserved C-terminal domain (TBP$_C$ and TFIIB$_C$), which consists of two direct repeats. The TBP N-terminal domain (TBP$_N$) is very poorly conserved and largely dispensable for function, while the TFIIB N-terminal domain (TFIIB$_N$) is highly conserved and essential. (B, C) TBP$_C$ bound to DNA. The two repeats are color coded as in (A). In (C), a magnified view of a phenylalanine sidechain from the C-terminal stirrup and two adjacent basepairs in the TATA box is included. The Phe side chain wedges apart the basepairs introducing a kink in the DNA. (D, E) The TBP$_C$/TFIIB$_C$/DNA complex. The TBP$_C$ and TFIIB$_C$ repeats are color coded as in (A). The position of the B recognition element (BRE), which binds the recognition helix of a helix-turn-helix motif in TFIIB$_C$, is indicated in (D).

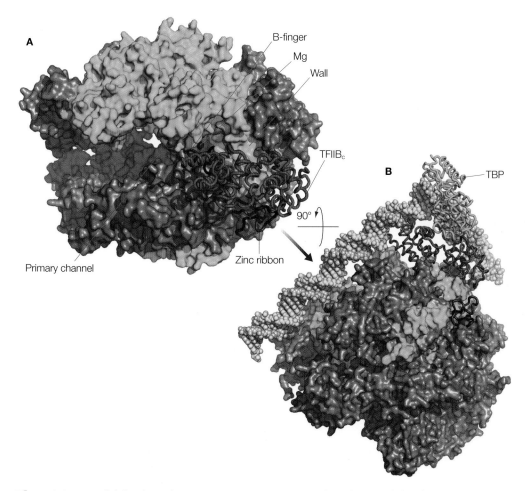

A

B-finger
Mg
Wall

TFIIB$_C$

B

TBP

90°

Primary channel

Zinc ribbon

Plate 3.2 *A model for the Pol II/TBP/TFIIB/promoter complex.* This model is based in part on X-ray crystal structures of the TBP$_C$/TFIIB$_C$/DNA complex and of the TFIIB$_N$/Pol II complex, and, in part, on chemical probing experiments. (A) This view shows only TFIIB and Pol II. Pol II is shown in a surface view. The wall (equivalent to the β-flap in the bacterial core polymerase) is green and the rest of the Rpb2 subunit is cyan, while the remainder of Pol II is tan. TFIIB is shown as a backbone trace, with TFIIB$_C$ colored magenta and TFIIB$_N$ colored blue. The active site Mg is shown as a blue sphere, while the zinc of the TFIIB zinc ribbon is shown as a red sphere. The zinc ribbon is bound to the dock at the tip of the wall, placing it close to the site where RNA is thought to exit the polymerase during elongation. The TFIIB polypeptide chain following the zinc ribbon disappears under the wall and emerges in the active site close to the Mg ion. The TFIIB linker region is thought to lead out of the active site through the primary channel where it connects with the TFIIB$_C$ domain. (B) In this rendering, TBP$_C$ and the promoter DNA have been added to the view. The coloring of Pol II and TFIIB is as in (A). TBP$_C$ is shown in cyan as a backbone trace. The DNA is shown in spacefilling view. The severe bend in the DNA induced by TBP$_C$ together with the binding of TFIIB$_C$ both upstream and downstream of the TATA box serves to bend the DNA around the preinitiation complex. This model was rendered using coordinates provided by Steven Hahn (Fred Hutchinson Cancer Research Center), from Chen, H.T. and Hahn, S. (2004) *Cell*, **119**, 169–180, and is based on the following structures from the Protein Data Bank: PDB id 1SFO, PDB id 1C9B, and PDB id1R5U.

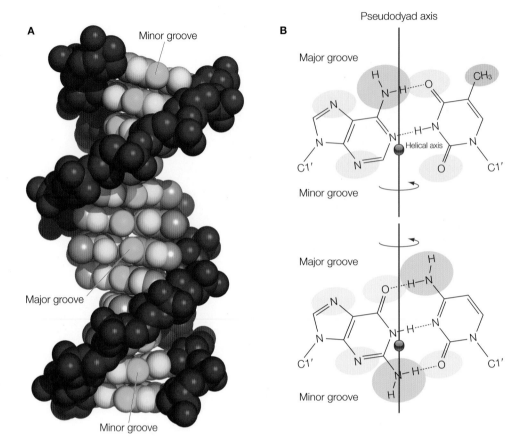

Plate 4.1 *H-bond donor and acceptor geometry in the DNA grooves.* (A) A side view of a spacefilling model of the DNA double helix (PDB id 1BNA). The sugar phosphate backbone atoms are colored red. Most of the base atoms are colored gray, with the exception of the H-bond donors (green), H-bond acceptors (yellow), and methyl groups (pink) on the minor and major groove edges of the bases. Each sequence presents a different geometric array of the H-bond donors and acceptors allowing sequence-specific DNA-binding proteins to recognize the different sequences. (B) The AT and GC basepairs. H-bond donors and acceptors, as well as methyl groups are shaded as in (A). The vertical line represents the pseudodyad axis that runs through each basepair and intersects the helical axis (circle). Rotation of the basepair by 180° about the pseudodyad axis results in the two C1′ carbon atoms exchanging positions with one another. Note that the pattern of H-bond donors and acceptors in the minor groove does not change upon rotation by 180° about the pseudodyad axis. As a result, the minor groove contains less sequence information than does the major groove.

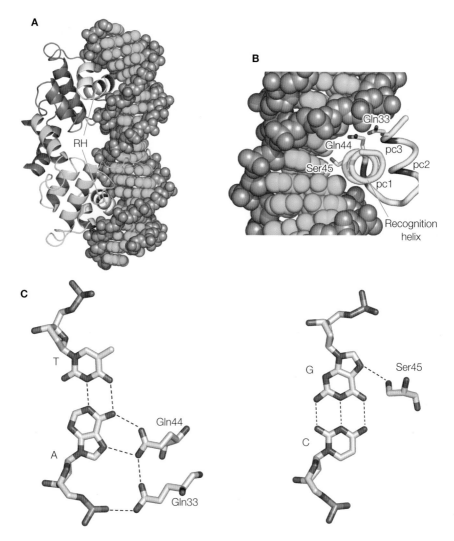

Plate 4.2 *Structure of the λcI protein DNA-binding domain* (PDB id 1LMB). (A) A λcI DNA-binding domain dimer bound to a λ operator site. This crystal structure contains just the N-terminal ~90 amino acids of the protein bound to a single operator site. The protein domain consists of a five α-helix bundle. Helices 2 and 3 in this bundle constitute a helix-turn-helix (HTH) motif. The two monomers that comprise the dimer are shown in blue and green, with the exception of the HTH motifs which are colored yellow. RH, recognition helix. (B) A close-up view of one of the HTH motifs bound to an operator half-site. The amino acid residues that are altered in the known λcI pc mutations (pc1, pc2, and pc3) are colored magenta. Two side chains (Gln44 and Gln33) that make contacts required for sequence-specific recognition are shown. (C) Left: An illustration of the H-bond contacts between Gln44, Gln33, and an AT basepair in the operator. Gln44, which is in the recognition helix, makes two H-bond contacts with the edge of the adenine residue. Gln33 which is in the first helix of the HTH motif makes H-bond contact to both Gln44 and a backbone phosphate helping to brace the recognition helix in the correct position to allow for sequence-specific binding to DNA. Right: H-bond contact between Ser45 and a GC basepair in the operator.

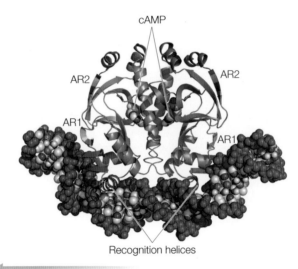

cAMP

AR2   AR2

AR1   AR1

Recognition helices

```
GAAAAGTGTGACAT ATGTCACACTTTTC
CTTTTCACACTGTA TACAGTGTGAAAAG
```

Plate 4.3 *The CRP dimer bound to its recognition element.* The sequence of the recognition element is shown at the bottom. It forms an inverted repeat, the two halves of which are recognized by the two different recognition helices in the CRP dimer (PDB id 1CGP). The two subunits in the dimer are colored blue and green. Residues that are altered in the known CRP pc mutations are colored yellow or red. The yellow residues define AR1, which is the region that contacts the RNA polymerase α subunit C-terminal domain, while the red residues define AR2, which is the region that contacts the RNA polymerase α subunit N-terminal domain. One molecule of cAMP (spacefilling view) is bound to each subunit.

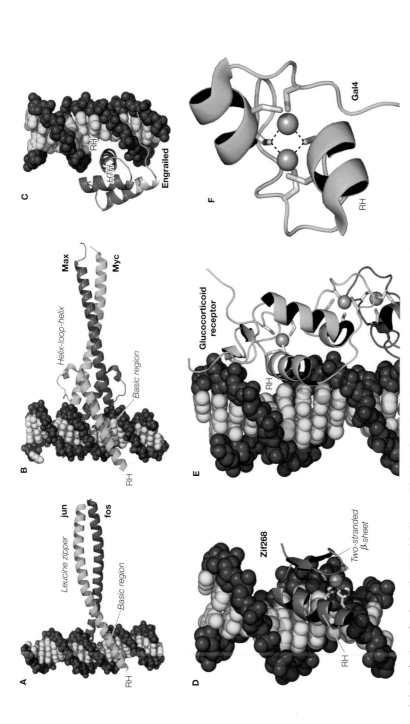

Plate 4.4 *A selection of eukaryotic DNA-binding motifs.* (A) The c-fos:c-jun heterodimer contains a bZip domain (PDB id 1FOS). (B) The Myc:Max heterodimer contains a bHLH domain (PDB id 1NKP). (C) Engrailed contains a homeodomain. The second (magenta) and third (orange) α-helices constitute a helix-turn-helix (HTH motif) (PDB id 1HDD). (D) Zif268 contains a Cys_2His_2 zinc finger domain. The two Cys and two His side chains that coordinate the single divalent zinc ion (magenta sphere) are shown (PDB id 1ZAA). (E) The glucocorticoid receptor contains a Cys_4Cys_4 zinc-binding domain. Glucocorticoid receptor binds DNA as a symmetric homodimer. The two subunits are colored cyan and magenta (only a portion of the magenta subunit is in view). The eight Cys side chains that coordinate the two zinc ions (pinkish spheres) are shown (PDB id 1R4O). (F) Gal4 contains a Cys_6 zinc-binding domain, sometimes termed a binuclear cluster. The six Cys side chains that coordinate the two zinc ions (pinkish spheres) are shown. The two central Cys side chains coordinate both zinc ions so that each zinc ion is coordinated by four ligands (PDB id 1D66). In (A–E) the protein is shown bound to DNA; in (F) the DNA is omitted. Each motif contains a recognition helix (RH) that binds in the major groove of the DNA. Only the DNA-binding domain of each protein is present in the structure.

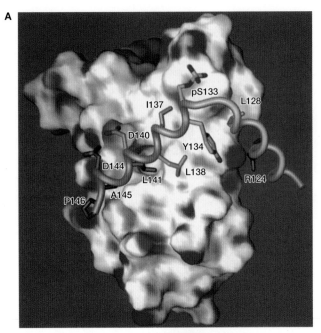

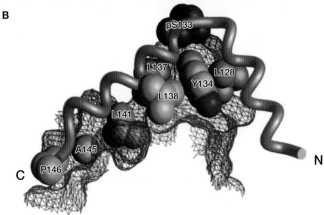

Plate 4.5 *Binding of the pKID activation domain to CBP.* (A) Structure
determined by NMR of the pKID activation domain from CREB bound
to a domain in CBP termed the KIX domain. The KIX domain is shown
in surface view with positively charged regions in blue and negatively
charged regions in red. The pKID domain binds in a neutral groove on
the surface of the KIX domain. (B) The hydrophobic interaction between
pKID and KIX. The hydrophobic side chains in pKID as well as the
phosphorylated Ser133 side chain are shown as spacefilling spheres.
The surface of a part of KIX is shown as a mesh to illustrate the shape
complementarity between KIX and pKID. From Radhakrishnan, I. et al.
(1997) *Cell*, **91**, 741–752, with permission from Elsevier.

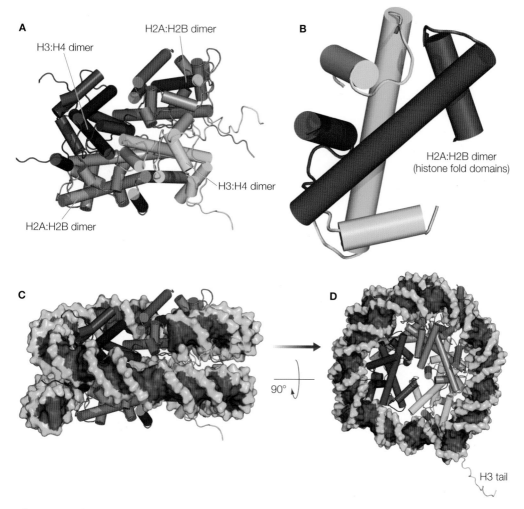

Plate 5.1 *The nucleosome core particle* (PDB id 1AOI). (A) The histone octamer. This consists of two H3:H4 dimers and two H2A:H2B dimers. The H3:H4 dimers associate to form a tetramer, which is flanked on either side by an H2A:H2B dimer. Each histone contains a globular domain consisting of a series of α-helices linked by turns. In this cartoon view, the α-helices are depicted as cylinders. In addition to these globular domains, the histones contain extended N-terminal tails; H2A and H2B also contain C-terminal tails. These extended tails are mobile and therefore mostly invisible in this structure. (B) An H2A:H2B histone fold pair. At the core of each histone dimer is a histone fold pair generated by the association of two histone fold domains (HFDs). Each HFD consists of three α-helices and thus the histone fold pair is a bundle of six helices. (C, D) The atomic resolution structure of the complete nucleosome core particle: (C) side view, (D) view down the superhelical axis. The histone octamer is wrapped by 147 bp of DNA forming two left-handed superhelical turns.

A

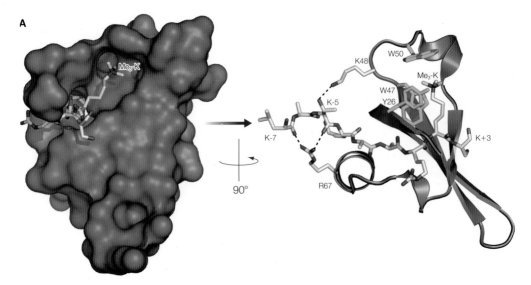

Plate 5.2A *Chromodomain and bromodomain structure.* (A) The Polycomb chromodomain complexed to a portion of the histone H3 N-terminal tail containing trimethyl-lysine 27 (PDB id 1PDQ). The image on the left shows a surface view of the chromodomain, while the image on the right shows a cartoon view of the chromodomain. The histone H3 peptide is depicted in wireframe view. The residues in this peptide are numbered relative to the trimethyl-lysine (Me₃-K), e.g., K-5 is the residue 5 positions N-terminal to the trimethyl-lysine. The trimethyl-lysine sits in a pocket lined with several aromatic side chains (Y26, W47, W50; shown as green wireframes on the right). The portion of the peptide N-terminal to the trimethyl-lysine threads through a channel that is only open to the solvent at its ends. H-bond contacts between chromodomain side chains (shown as wireframes – CPK colors; see Box 1.2) and atoms in the peptide provide specificity to the interaction. These include an H-bond between the Lys48 side chain in the chromodomain and the Thr side chain at position K-5 in the peptide; an H-bond between the Arg67 side chain in the chromodomain and the Thr carbonyl oxygen at position K-5 in the peptide; and an H-bond between the Arg67 side chain in the chromodomain and the leucine carbonyl oxygen at position K-7 in the peptide.

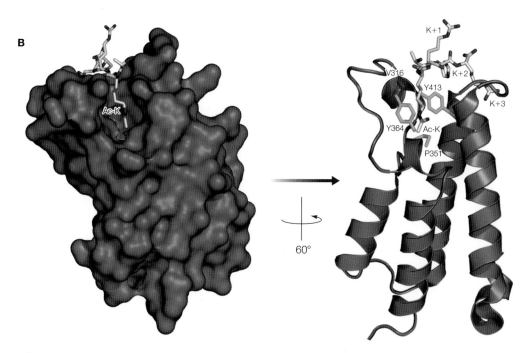

Plate 5.2B (*continued*) *Chromodomain and bromodomain structure.* (B) The GCN5 bromodomain complexed to a portion of the histone H4 N-terminal tail containing acetyl-lysine 16 (PDB id 1E6I). The image on the left shows a surface view of the bromodomain, while the image on the right shows a cartoon view of the bromodomain. The histone H4 peptide is depicted in wireframe view. The residues in this peptide are numbered relative to the acetyl-lysine (Ac-K), e.g., K+3 is the residue 3 positions C-terminal to the acetyl-lysine. The acetyl-lysine sits in a pocket lined with hydrophobic side chains (V316, P351, Y364, Y413; shown as green wireframes on the right).

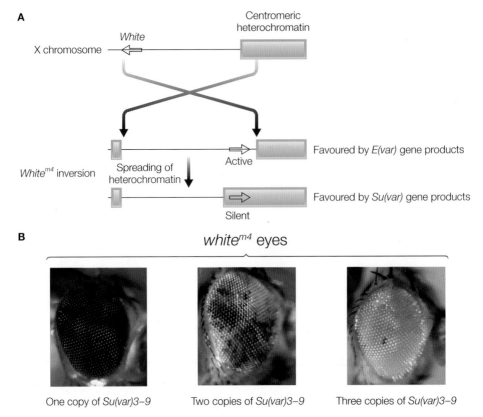

A

White

Centromeric
heterochromatin

X chromosome ⟵

White^{m4} inversion

Spreading of
heterochromatin ↓

Active

Favoured by *E(var)* gene products

Silent

Favoured by *Su(var)* gene products

B

white^{m4} eyes

One copy of *Su(var)3–9*   Two copies of *Su(var)3–9*   Three copies of *Su(var)3–9*

Plate 6.1 *Position effect variegation.* (A) The *white^{m4}* inversion. The *white* gene is located close to the left end of the wild-type *Drosophila* X chromosome. This chromosome has only one "arm" as the centromere is located at one end. An inversion (called *white^{m4}*) that reverses the direction of most of this arm places *white* right next to the centromeric heterochromatin. If the heterochromatin spreads to include *white*, the gene falls silent. Formation and spreading of heterochromatin is favored by the products of the *Su(var)* genes and disfavored by the products of the *E(var)* genes. (B) *white^{m4}* fly eyes in wild type and *Su(var)3-9* mutant backgrounds. The *white^{m4}* eye consists of a mixture of white tissue, in which the *white* gene was inactive during eye development because of spreading of heterochromatin, and red tissue in which the *white* gene was active during eye development because of the failure of heterochromatin to spread. Large patches of red and white tissue are often observed. These patches result from the epigenetic inheritance of the silent or active state during the proliferation of the cells that eventually give rise to the eye. Mutations in *Su(var)* and *E(var)* genes alter the probability that *white* will be silenced by the spreading of heterochromatin. The eye shown in the middle panel is from a fly that contains two wild-type copies of each *Su(var)* and *E(var)* gene and thus shows the characteristic mixture of white and red tissue. The eye shown in the left panel is from a fly in which one of the two copies of the *Su(var)3-9* gene has been deleted. *Su(var)3-9* encodes a histone methyltransferase required for heterochromatin formation and thus we see reduced heterochromatic silencing in this eye, which is therefore all red. The eye shown in the right panel is from a fly containing an extra wild-type copy of the *Su(var)3-9* gene. As a result, we observed increased heterochromatic silencing in the eye, which is therefore nearly all white. The phenotype resulting from an extra copy of a *Su(var)* gene is very similar to that resulting from the inactivation of *E(var)* genes. Photographs courtesy of Gunter Reuter (Institut fur Genetik, Biologicum, Martin-Luther-Universitat), with permission from Elsevier.

Plate 6.2 *A calico cat*. The patches of orange and black fur result from random inactivation of the X chromosome during early embryogenesis. The areas of white fur result from a process unrelated to X chromosome inactivation. Photograph courtesy of Marisa Duke.

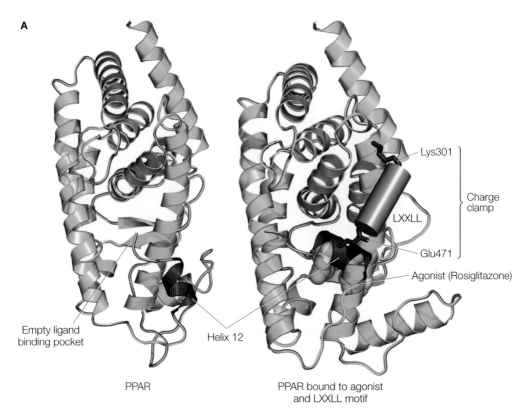

A

Lys301

Charge clamp

LXXLL

Glu471

Agonist (Rosiglitazone)

Empty ligand binding pocket

Helix 12

PPAR

PPAR bound to agonist and LXXLL motif

Plate 7.1A *Allosteric regulation of nuclear receptor ligand-binding domains.* (A) The X-ray crystal structure of PPAR in the absence of ligand (left; PDB id 1PRG) and in the presence of a synthetic ligand termed Rosiglitazone known to function as an agonist (right; PDB id 2PRG). The ligand-bound receptor is also bound to an LXXLL motif common to many coactivators that bind nuclear receptors. PPAR (ribbon diagram) is green with the exception of helix 12, which is blue. The ligand (spacefilling model) is yellow. The LXXLL motif (shown as a cylinder) is magenta. Note that in the presence of agonist, helix 12 moves up over the binding pocket sealing off the ligand from the solvent. In the ligand-bound conformation, Glu471 on helix 12 and Lys301 form a "charge clamp" which binds tightly to the LXXLL motif.

B

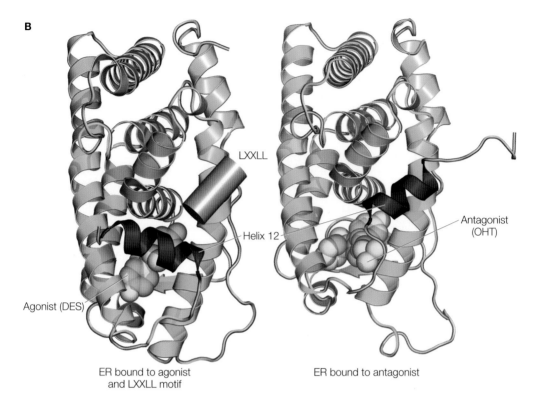

LXXLL

Helix 12

Antagonist (OHT)

Agonist (DES)

ER bound to agonist and LXXLL motif

ER bound to antagonist

Plate 7.1B (*continued*) *Allosteric regulation of nuclear receptor ligand-binding domains.* (B) X-ray crystal structure of the estrogen receptor (ER) bound to a synthetic ER agonist known as diethylstilbestrol (DES) (left; PDB id 3ERD) and to a synthetic ER antagonist known as 4-hydroxytamoxifen (OHT) (right; PDB id 3ERT). The DES-bound receptor is also bound to an LXXLL coactivator motif. The antagonist binds in the same binding pocket as the agonist. However, the antagonist is too bulky to allow helix 12 to assume its normal position. Instead helix 12 assumes the position normally occupied by the LXXLL coactivator motif, thus blocking coactivator binding.

PROTEIN DATA BANK STRUCTURE REFERENCES FOR COLOR PLATES

Plate B1.1
PDB id 1ZAA: Pavletich, N.P. and Pabo, C.O. (1991) Zinc finger–DNA recognition: crystal structure of a Zif268–DNA complex at 2.1 A. *Science*, **252**, 809–817.

Plate 2.1
PDB id 1HQM: Minakhin, L., Bhagat, S., Brunning, A., Campbell, E.A., Darst, S.A., Ebright, R.H. and Severinov, K. (2001) Bacterial RNA polymerase subunit omega and eukaryotic RNA polymerase subunit RPB6 are sequence, structural, and functional homologs and promote RNA polymerase assembly. *Proc Natl Acad Sci USA*, **98**, 892–897.

Plate 2.2
PDB id 1HQM: Minakhin, L., Bhagat, S., Brunning, A., Campbell, E.A., Darst, S.A., Ebright, R.H., Severinov, K. (2001) Bacterial RNA polymerase subunit omega and eukaryotic RNA polymerase subunit RPB6 are sequence, structural, and functional homologs and promote RNA polymerase assembly. *Proc Natl Acad Sci USA*, **98**, 892–897.
PDB id 1I6H: Gnatt, A.L., Cramer, P., Fu, J., Bushnell, D.A. and Kornberg, R.D. (2001) Structural basis of transcription: an RNA polymerase II elongation complex at 3.3 A resolution. *Science*, **292**, 1876–1882.

Plate 3.1
PDB id 1D3U: Littlefield, O., Korkhin, Y. and Sigler, P.B. (1999) The structural basis for the oriented assembly of a TBP/TFB/promoter complex. *Proc Natl Acad Sci USA*, **96**, 13668–13673.

Plate 3.2
PDB id 1SFO: Westover, K.D., Bushnell, D.A. and Kornberg, R.D. (2004) Structural basis of transcription: separation of RNA from DNA by RNA polymerase II. *Science*, **303**, 1014–1016.
PDB id 1C9B: Tsai, F.T. and Sigler, P.B. (2000) Structural basis of preinitiation complex assembly on human pol II promoters. *Embo J*, **19**, 25–36.
PDB id 1R5U: Bushnell, D.A., Westover, K.D., Davis, R. and Kornberg, R.D. (2004) Structural basis of transcription: an RNA polymerase II-TFIIB cocrystal at 4.5 angstroms. *Science*, **303**, 983–988.

Plate 4.1
PDB id 1BNA: Drew, H.R., Wing, R.M., Takano, T., Broka, C., Tanaka, S., Itakura, K. and Dickerson, R.E. (1981) Structure of a B-DNA dodecamer: conformation and dynamics. *Proc Natl Acad Sci USA*, **78**, 2179–2183.

Plate 4.2
PDB id 1LMB: Beamer, L.J. and Pabo, C.O. (1992) Refined 1.8 A crystal structure of the lambda repressor–operator complex. *J Mol Biol*, **227**, 177–196.

Plate 4.3
PDB id 1CGP: Schultz, S.C., Shields, G.C. and Steitz, T.A. (1991) Crystal structure of a CAP–DNA complex: the DNA is bent by 90 degrees. *Science*, **253**, 1001–1007.

Plate 4.4
PDB id 1FOS: Glover, J.N. and Harrison, S.C. (1995) Crystal structure of the heterodimeric bZIP transcription factor c-Fos-c-Jun bound to DNA. *Nature*, **373**, 257–261.
PDB id 1NKP: Nair, S.K. and Burley, S.K. (2003) X-ray structures of Myc-Max and Mad-Max recognizing DNA: molecular bases of regulation by proto-oncogenic transcription factors. *Cell (Cambridge, Mass)*, **112**, 193–205.
PDB id 1HDD: Kissinger, C.R., Liu, B.S., Martin-Blanco, E., Kornberg, T.B. and Pabo, C.O. (1990) Crystal structure of an engrailed homeodomain–DNA complex at 2.8 A resolution: a framework for understanding homeo-domain–DNA interactions. *Cell*, **63**, 579–590.
PDB id 1ZAA: Pavletich, N.P. and Pabo, C.O. (1991) Zinc finger–DNA recognition: crystal structure of a Zif268–DNA complex at 2.1 A. *Science*, **252**, 809–817.
PDB id 1R4O: Luisi, B.F., Xu, W.X., Otwinowski, Z., Freedman, L.P., Yamamoto, K.R. and Sigler, P.B. (1991) Crystallographic analysis of the interaction of the glucocorticoid receptor with DNA. *Nature*, **352**, 497–505.
PDB id 1D66: Marmorstein, R., Carey, M., Ptashne, M. and Harrison, S.C. (1992) DNA recognition by GAL4: structure of a protein–DNA complex. *Nature*, **356**, 408–414.

Plate 5.1
PDB id# 1AOI: Luger, K., Mader, A.W., Richmond, R.K., Sargent, D.F. and Richmond, T.J. (1997) Crystal structure of the nucleosome core particle at 2.8 A resolution. *Nature*, **389**, 251–260.

Plate 5.2
PDB id 1PDQ: Fischle, W., Wang, Y., Jacobs, S.A., Kim, Y., Allis, C.D. and Khorasanizadeh, S. (2003) Molecular basis for the discrimination of repressive methyl-lysine marks in histone H3 by Polycomb and HP1 chromodomains. *Genes Dev*, **17**, 1870–1881.
PBD id 1E6I: Owen, D.J., Ornaghi, P., Yang, J.C., Lowe, N., Evans, P.R., Ballario, P., Neuhaus, D., Filetici, P. and Travers, A.A. (2000) The structural basis for the recognition of acetylated histone H4 by the bromodo-main of histone acetyltransferase gcn5p. *Embo J*, **19**, 6141–6149.

Plate 7.1
PDB id 1PRG and **PDB id 2PRG**: Nolte, R.T., Wisely, G.B., Westin, S., Cobb, J.E., Lambert, M.H., Kurokawa, R., Rosenfeld, M.G., Willson, T.M., Glass, C.K. and Milburn, M.V. (1998) Ligand binding and co-activator assembly of the peroxisome proliferator-activated receptor-gamma. *Nature*, **395**, 137–143.
PDB id 3ERD and **PDB id 3ERT**: Shiau, A.K., Barstad, D., Loria, P.M., Cheng, L., Kushner, P.J., Agard, D.A. and Greene, G.L. (1998) The structural basis of estrogen receptor/coactivator recognition and the antagonism of this interaction by tamoxifen. *Cell*, **95**, 927–937.

Box 4.3

The abortive initiation assay

The process of transcriptional initiation involves many steps, including recruitment of RNA polymerase (closed complex formation), the melting of the DNA around the promoter (open complex formation), the formation of the first few phosphodiester bonds while the polymerase is still bound to the promoter (abortive initiation), and the release of the polymerase from the promoter (promoter clearance) to form a stable ternary elongation complex. An understanding of how activators facilitate this overall process requires techniques for measuring the physical parameters (equilibrium binding constants and rate constants) associated with the individual steps. The abortive initiation assay is one such technique. This procedure is designed to look at the first two steps in initiation: closed complex formation and open complex formation.

    In the abortive initiation assay, one starts by mixing together a promoter, RNA polymerase, and a partial set of nucleotide substrates. The nucleotides included in the reaction are selected so that the RNA polymerase cannot synthesize a transcript of more than 2–10 nucleotides in length before it stalls due to the absence of a needed nucleotide. For example, in the case of the *E. coli* lac promoter (Figure B4.3A), the abortive initiation assay mixture includes just ATP and UTP (omitting GTP and CTP). Since the fifth nucleotide in the lac mRNA is a G, the polymerase will only be able to synthesize transcripts up to four nucleotides in length. Recall that formation of the stable ternary elongation complex by *E. coli* RNA polymerase only occurs when the transcript is about 10 nucleotides in length.

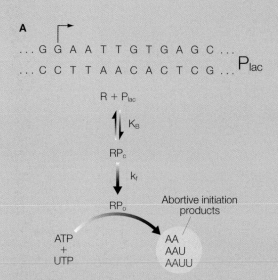

Figure B4.3 *Generating and analyzing abortive initiation data.* (A) Abortive initiation at the lac promoter. The sequence at the top is that of the *E. coli* lac promoter with the arrow indicating the transcriptional start site.

As a result, the polymerase in this reaction mixture will be stuck in the abortive initiation mode continuously synthesizing and releasing abortive transcripts of up to four nucleotides in length, recycling back to the open complex after releasing each abortive transcript.

Under the circumstances described above, the rate at which abortive initiation products accumulate will be directly proportional to the concentration of the open complex. A plot of the concentration of abortive initiation products versus time will show a "lag" phase followed by a linear phase (Figure B4.3B). The length of this lag phase, which is termed τ_{obs}, is dictated by the time required to form the open complex. The length of the lag depends on the stability of the closed complex as given by the equilibrium binding constant K_B; and by the speed at which the closed complex is converted to the open complex as given by the rate constant k_f (Figure B4.3A). By solving the rate equation for open complex formation, one can generate an equation relating τ_{obs} to the RNA polymerase concentration. It turns out that a graph of τ_{obs} versus the reciprocal of the RNA polymerase concentration gives a straight line and that the slope and intercept of this line can be analyzed to yield K_B and k_f (Figure B4.3C).

This is the approach that was used to determine the values for K_B and k_f given in Table 4.1. By measuring these parameters in the absence and presence of activators it was possible to show that CRP accelerates transcription of the lac promoter by increasing K_B, while the λcI promoter accelerates transcription of P_{RM} by increasing k_f.

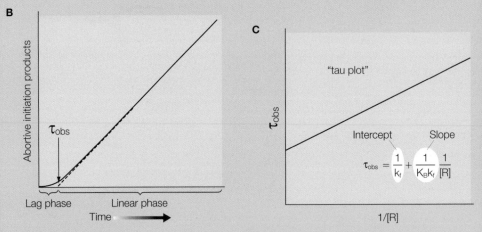

Figure B4.3 (*continued*) *Generating and analyzing abortive initiation data.* (B) A graph of the concentration of abortive initiation products versus time. The graph exhibits a lag before reaching a linear phase. Extrapolation of the linear portion of the curve to the *x*-axis yields τ_{obs}, the lag time. (C) A tau plot. A graph of τ_{obs} versus the reciprocal of the RNA polymerase concentration will often give a straight line with the equation shown. K_B and k_f can then be determined from the slope and intercept of the line.

Table 4.1

Effect of λcI and cAMP receptor protein (CRP) on initiation kinetics[1]

Promoter	Activator	K_B (M^{-1})	k_f (s^{-1})
P_{lac}	None	1.5×10^7	3×10^{-3}
	CRP	$\geq 2 \times 10^8$	3×10^{-3}
P_{RM}	None	9.9×10^6	7.0×10^{-4}
	λcI	9.2×10^6	7.8×10^{-3}

[1] Values were determined in abortive initiation assays (see Box 4.3). Data on P_{RM} are from Hawley, D.K. and McClure, W.R. (1982) Mechanism of activation of transcription initiation from the lambda PRM promoter. *J Mol Biol*, **157**, 493–525. Data on P_{lac} are from Malan, T.P., Kolb, A., Buc, H. and McClure, W.R. (1984) Mechanism of CRP-cAMP activation of lac operon transcription initiation activation of the P1 promoter. *J Mol Biol*, **180**, 881–909.

the rate of lac promoter open complex formation demonstrate that CRP stabilizes the closed complex relative to the free promoter by about a factor of 10, while having little or no effect on the rate at which the closed complex is converted to the open complex. In contrast, abortive initiation assays carried out on P_{RM} demonstrate that λcI accelerates the conversion of the closed complex to the open complex by a factor of about 10, while having little or no effect on the stability of the closed complex. In other words, CRP increases K_B, while λcI increases k_f (Table 4.1).

Recruitment versus allostery

How do the results of these abortive initiation assays jibe with the studies showing that activation requires direct contacts between activators and RNA polymerase? Explanation of the CRP result is straightforward, since the increased stability of the lac promoter closed complex (reflected by the increase in K_B) in the presence of CRP can be readily attributed to a stabilizing contact between AR1 and the RNA polymerase α subunit. This type of activation, in which an activator favors the binding of RNA polymerase to DNA, is often described as "recruitment", i.e., CRP is said to recruit RNA polymerase to the promoter. Recruitment is a general mechanism that can explain many examples of transcriptional activation in both bacteria and eukaryotes.

In the case of activation by λcI, things are more complicated. The abortive initiation assays indicate that this activator has no effect on the stability of the closed complex relative to the free promoter, but instead stimulates open complex formation. How might an interaction between λcI and σ[70] favor promoter opening? Perhaps the most obvious possibility is that the activator serves as an allosteric regulator of RNA polymerase. In allosteric theory, enzymes are considered to exist in multiple conformational states that have different affinities for various ligands. If a ligand binds to conformation A more tightly than to conformation B, conformation A will be favored by the ligand. If conformation A is the active state, then the ligand is an allosteric activator of the enzyme, while if conformation A is the inactive state, then the ligand is an allosteric inhibitor.

We can easily apply these ideas to RNA polymerase by considering the closed complex to be the inactive state and the open complex to be the active state and by considering the activator to be the allosteric ligand. As we saw in Chapter 2, conversion of the closed complex to the open complex involves major changes in the conformation of both the DNA and the polymerase. It is therefore easy to imagine that an activator would bind to one of these conformations and not the other. If binding of λcI to the polymerase only occurs when polymerase is in the open complex, then promoter opening will be favored by the presence of λcI, and λcI will function as an allosteric activator.

4.3 EUKARYOTIC ACTIVATORS AND THEIR TARGETS

4.3.1 The modular nature of eukaryotic activators

The preceding discussion of bacterial activators indicates that they require at least two interaction surfaces: a DNA interaction surface for binding to cis-regulatory elements and a protein interaction surface for contacting a component of the transcriptional machinery. In bacteria, these two functions are often integrated into a single globular domain, with a particularly striking example provided by λcI, in which a single HTH motif contains both DNA-binding and RNA polymerase-binding interfaces.

In contrast, eukaryotic activators commonly exhibit modular structures in which DNA binding and transcriptional activation are conferred by independently folding protein domains. As a result of this modular structure, it is often possible to link the activation domain from one factor to the DNA-binding domain from another factor, creating a chimeric factor with a novel combination of DNA-binding and transcriptional activation properties. Indeed, one of the most intensively studied eukaryotic activators, called Gal4-VP16, is not a naturally occurring activator, but a chimera created in the lab. VP16 is a herpes simplex virus-encoded activator that contains a particularly potent activation domain. However, VP16 does not possess a DNA-binding domain and is recruited to viral promoters by DNA-bound transcription factors encoded by the host genome. To study activation by VP16 in yeast, which lacks the DNA-bound transcription factors that recruit VP16, it was necessary to fuse the VP16 activation domain to the well-defined DNA-binding domain from the yeast activator

Gal4. The resulting chimeric activator, Gal4-VP16, could then bind and activate genes containing Gal4-binding sites.

Diversity in the structure of eukaryotic DNA-binding domains

Both of the bacterial activators discussed earlier in this chapter, CRP and λcI, recognize DNA using an interface known as the helix-turn-helix motif. As was explained above, the HTH motif provides a means of precisely positioning an α-helix in the major groove where it can make well-defined contacts with the major groove edges of the basepairs. In bacteria, the HTH motif is by far the most common motif for sequence-specific DNA recognition in transcriptional regulation. However, the situation in eukaryotes is more complex – there are many different families of DNA-binding motifs in activator proteins, and no single family is dominant. Most of these motifs represent different solutions to the basic problem of how to position an α-helix (the recognition helix) in the major groove. This is a good way to recognize specific DNA sequences because the α-helix and the major groove have roughly complementary shapes, because the rigidity of the α-helix enables it to present side chains to the major groove in a well-defined manner, and because the complex array of H-bond donors and acceptors in the major groove makes it a rich source of sequence information. This section will summarize the properties of a few of those motifs including the basic leucine zipper (bZip) domain, the basic helix-loop-helix (bHLH) domain, the homeodomain, and the zinc-binding domains. This is by no means an exhaustive list.

The bZip and bHLH domains. Perhaps the simplest of all DNA-binding motifs is the bZip domain (Plate 4.4A), which is exemplified by the intensively studied yeast transcription factor Gcn4, as well as the closely related mammalian factors AP-1 and CREB (discussed further later in this chapter). The bZip domain consists of nothing more than an approximately 40 amino acid-long α-helix. The N-terminal third of this helix, which contains multiple positively charged side chains, is termed the basic region, and constitutes the DNA-binding interface. The C-terminal two-thirds of the helix is termed the leucine zipper and constitutes a protein dimerization interface. The leucine zipper is amphipathic, meaning that it has a hydrophilic face and a hydrophobic face. The hydrophobic faces on two copies of the leucine zipper bind to one another to bring the two bZip domains together in a parallel orientation. This brings together two basic regions that can then bind to the DNA by passing through the major groove on opposite sides of the double helix forming a "scissors grip" on the DNA.

Somewhat more complex than the bZip domain is the bHLH domain (Plate 4.4B), which is found in many factors that regulate growth and differentiation, such as the Myc:Max heterodimer (discussed in Chapter 5).

Like the bZip domain, the bHLH domain also contains a positively charged α-helix (the basic region) that passes through the major groove and serves as the DNA-binding interface, which is directly connected to a dimerization interface. In this case, the dimerization interface consists of two α-helices connected by a loop (hence the name helix-loop-helix). The four α-helices in the two subunits of the dimeric helix-loop-helix motif pack together in a four-helix bundle.

The homeodomain. Eukaryotic genomes also encode many DNA-binding motifs that are structurally and in some cases evolutionarily related to the bacterial HTH motif. Perhaps the most important motif of this type is the homeodomain (Plate 4.4C). The homeotic gene complex (see Chapter 6), which is found in all segmental organisms, including insects and vertebrates, encodes a family of homeodomain-containing transcription factors such as Antp and Ubx that control segment identity (discussed extensively in Chapters 6 and 7). Studies of these genes in *Drosophila* demonstrate that mutations in genes encoding single transcription factors can have profound effects on development (e.g., legs growing out of the head), greatly enhancing our understanding of the connection between molecular evolution and development.

Zinc-containing DNA-binding domains. A huge number of eukaryotic DNA-binding domains are stabilized by interactions with zinc. In all of these domains, one or more zinc ions form tetrahedral coordination complexes with Cys and/or His side chains (Plate 4.4D–F). These domains always include a recognition helix that packs into the major groove of the recognition element. One very large family of zinc-containing motifs are the Cys_2His_2 zinc finger motifs (Plate 4.4D) found in many transcription factors including *Drosophila* Kruppel and human Sp1. In these motifs, two Cys and two His side chains complex a single zinc ion. The motif folds into a two-stranded antiparallel β-sheet followed by an α-helix; the zinc ion bridges these two secondary structural elements. The α-helix serves as a recognition helix: it passes through the major groove and contacts three successive basepairs in the recognition element.

Another family of zinc-binding domains, the Cys_4Cys_4 domains (Plate 4.4E), include the DNA-binding domains of the nuclear receptors such as glucocorticoid receptor, estrogen receptor, retinoic acid receptor, thyroid hormone receptor, and vitamin D receptor. These receptors each comprise simple, wholly self-contained signaling systems. The signaling agents with which they interact (steroids, retinoids, etc.) are lipid-soluble compounds that freely cross the plasma membrane to bind the intracellular receptors. The binding of the ligand to its receptor alters the properties of the receptor allowing it to activate transcription of selected target genes. The DNA-binding domains in these factors each contain eight cysteine side chains involved in two tetrahedral coordination complexes with two zinc ions. The DNA-binding and activation properties of nuclear receptors will be discussed extensively in Chapter 7.

Yet another zinc-containing DNA-binding motif is the Zn_2Cys_6 binuclear cluster (Plate 4.4F). These domains are found only in fungi and include the well-studied yeast transcriptional activator Gal4, which, in the presence of galactose, activates genes required for galactose catabolism. Like the nuclear hormone receptors, this domain contains two zinc ions. However, in this case there are only six Cys residues involved in zinc coordination. Two of these Cys side chains form coordinate bonds to both zinc ions, thereby providing each zinc ion with the four ligands required to fill its tetrahedral coordination shell.

Activation domains

While DNA-binding domains in many eukaryotic factors are structurally and functionally well defined, the same cannot generally be said for activation domains. We know that these domains generally serve as protein –protein interaction domains, and, in many cases, the target proteins in the transcriptional machinery with which they interact are known. However, very little structural information about these domains is available.

Activation domains are often categorized according to their amino acid composition. Many activation domains (e.g., the Gal4 activation domain) are rich in the negatively charged amino acids, Asp and Glu, and are accordingly referred to as acidic activation domains. Other activation domains, for example, the Gln-rich activation domains in Sp1, have an abundance of other amino acids. Beyond this tendency to be rich in certain amino acids, however, activation domains are poorly conserved. This lack of sequence conservation and the paucity of activation domain structures in the literature lead to the conclusion that such domains are not usually well ordered. This idea is further re-enforced by deletion analysis of activation domains. These studies often show that activation domains contain multiple short segments with activation potential that work together in a more or less additive manner. This suggests that long range tertiary interactions are not required for activation domain function.

Inducing order in activation domains

If activation domains are so disordered, how can they mediate specific interactions with targets in the transcriptional machinery? The answer to this question may reside, at least in part, in the ability of the targets to induce order in the activation domains. An activation domain in the cAMP response element-binding protein (CREB) provides one of the best examples of this phenomenon. CREB is a bZip domain-containing activator protein that becomes phosphorylated by protein kinase A in response to the second messenger cAMP. The unique phosphoacceptor serine residue lies within a 60-residue long activation domain termed the kinase A inducible activation domain (KID). The phosphorylated form of KID (pKID) binds to a transcriptional coactivator termed CREB-binding protein (CBP), which functions as an adaptor between CREB and the general machinery and which also functions as a histone acetyltransferase. As we will see in the next chapter, histone acetylation is often linked to gene activation.

In the absence of CBP, pKID is a random coil. However, binding to CBP induces an approximately 25 amino acid-long segment of pKID to assume an ordered structure consisting of two α-helices. The second of these helices is amphipathic in nature, having one hydrophilic and one hydrophobic face. The hydrophobic face of this helix packs into a hydrophobic groove on the surface of CBP (Plate 4.5).

This strategy of using an amphipathic α-helix in an activation domain to pack into a hydrophobic groove in a target protein may be fairly general. For example, as we will see in Chapter 7, a hormone inducible activation domain found in many nuclear hormone receptors appears to function in just this way.

4.3.2 The Mediator: a special activator target in the eukaryotic transcriptional machinery

Activation in bacteria generally involves direct contacts between the DNA-bound activators and the transcriptional machinery leading to recruitment and/or allosteric activation of this machinery. Eukaryotic activators use these same mechanisms. However, the eukaryotic basal machinery has evolved to include additional targets for regulatory interactions that are absent from the bacterial machinery. Probably the most important activator target in the eukaryotic class II machinery is a protein complex termed the Mediator.

Biochemical and genetic identification of the Mediator

A minimal cell-free transcription system containing core RNA polymerase II and purified general transcription factors supports basal transcription from a core promoter. But activators such as Gal4-VP16 do not generally stimulate transcription in such a system. This suggests that activation must require one or more factors missing from the minimal system that mediate communication between activators and the basal machinery. The search for such factors led to the discovery of a protein complex termed the Mediator, which could exist as both an independent complex and in association with core Pol II. Addition of the Mediator to cell-free transcription systems allowed promoter activation by numerous activators including Gal4-VP16 and natural yeast activators such as Gcn4 and Gal4. The yeast Mediator contains about 22 subunits with a total mass well in excess of a million daltons (i.e., more than twice as large as the 12-subunit core RNA polymerase!) (Table 4.2). Although initially discovered in yeast, homologs of the Mediator are present in all eukaryotes. At least 19 of the 22 subunits in the yeast mediator have mammalian counterparts. In addition, the mammalian complex contains additional subunits not found in yeast (Table 4.2).

Evidence that the Mediator is really required for Pol II function in living cells came from genetic screens to find second site suppressor mutations that compensate for growth defects due to a mutation in Pol II. As discussed in both Chapters 2 and 3, in addition to the β′ homology region, the large subunit of Pol II contains a CTD consisting of multiple repeats of a seven amino acid sequence, with the number of repeats varying from species to species. Wild-type yeast Pol II contains 26 copies of the heptapeptide. While the CTD is essential for viability, a yeast strain carrying a truncated Pol II containing just 11 repeats is viable but exhibits numerous defects including cold sensitivity. When yeast carrying this cold-sensitive Pol II variant were mutagenized and screened for the loss of the cold sensitivity, a number of second site suppressor mutations were identified. These mutations were mapped to nine genes referred to as the suppressor of RNA polymerase B (SRB) genes. (Note that RNA polymerase B is an alternative name for RNA polymerase II.) Further analysis of these genes revealed that they encoded nine of the 22 subunits of the Mediator. The finding that mutations in Mediator subunits can suppress defects due to a mutation in Pol II strongly supports the idea that the Mediator is, in fact, required for Pol II function *in vivo*.

Consistent with the finding that Mediator mutations can compensate for a truncated CTD, the Mediator binds to Pol II in part through interactions with the CTD. The Mediator also binds directly to activators, for example, Gal4 and VP16. The implication of these findings is that the Mediator serves as an adaptor to facilitate communication between Pol II and activator proteins.

Mechanism of Mediator action

Just as different bacterial activators contact diverse surfaces in bacterial RNA polymerase, eukaryotic activators are able to latch on to many different parts of the Mediator. Electron microscopy shows that the Mediator contains three modules or domains termed the tail, middle, and head domains and wraps around core Pol II (Figure 4.2A). Contacts with activators have been mapped to all three Mediator domains. For example, Gal4 contacts both Med17 (which is in the head domain) and Med15 (tail domain), while nuclear receptors contact Med1 (middle domain).

As we saw in the discussion of the bacterial paradigms, contacts between activators and the bacterial RNA polymerase holoenzyme can

Table 4.2
Subunit composition of the Mediator of transcription

Name[1]	Length (amino acids): yeast, human[2]	Module	Comments
MED1	566, 1581	Middle	Nuclear receptor target
MED2	431, ?	Tail	
MED3	397, ?	Tail	
MED4	258, 270	Middle	Essential gene in yeast
MED5 (Nut1)	1132, ?	Middle	
MED6	295, 246	Head	Essential gene in yeast
MED7	222, 233	Middle	Essential gene in yeast
MED8	223, 289	Head	Essential gene in yeast
MED9 (Cse2)	149, 146	Middle	
MED10 (Nut2)	157, 135	Middle	Essential gene in yeast
MED11	131, 178	Head	Essential gene in yeast
MED12 (Srb8)	1427, 2212	Cdk8	
MED13 (Srb9)	1320, 2174	Cdk8	
MED14 (Rgr1)	1082, 1454	Tail	Essential gene in yeast
MED15 (Gal11)	1081, 767	Tail	KIX domain, Swi5, Gal4, Gcn4, Smad2/4 target
MED16 (Sin4)	974, 877	Tail	
MED17 (Srb4)	687, 651	Head	Essential gene in yeast, Gal4 target
MED18 (Srb5)	307, 208	Head	
MED19 (Rox3)	220, 194	Head	Essential gene in yeast
MED20 (Srb2)	210, 212	Head	
MED21 (Srb7)	140, 144	Middle	Essential gene in yeast, Tup1 target
MED22 (Srb6)	121, 200	Head	Essential gene in yeast
MED23	?, 1364	?	E1A, Elk1 target
MED24	?, 989	?	
MED25	?, 747	?	VP16 target
MED26	?, 600	?	
MED27	?, 311	?	
MED28	?, 178	?	
MED29	?, 221	?	
MED30	?, 178	?	
MED31 (Soh1)	127, 131	?	
CDK8 (Srb10)	555, 464	Cdk8	Tup1 target, CTD kinase
CycC (Srb11)	323, 283	Cdk8	Cyclin partner for CDK8

CTD, RNA polymerase II C-terminal domain.

[1]All the Mediator subunits have been given systematic names that apply across all species (Med1 to Med31, CDK8, CycC). Also given in parentheses are the original names of the yeast genes when they differ from the systematic names.

[2]The two numbers for each subunit represent the length in amino acid residues of the yeast and human orthologs. A question mark indicates that the yeast or human ortholog has not been identified and may not exist.

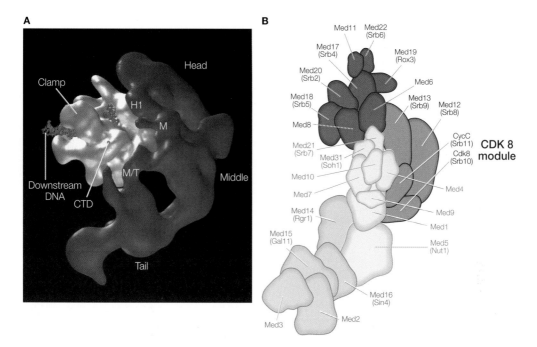

Figure 4.2 *Low resolution structure of the yeast Mediator.* (A) A three-dimensional reconstruction from electron microscopy showing yeast Pol II holoenzyme, including the core in white and the Mediator shaded in gray. Regions of the Mediator that contact the core are labeled H1 (head submodule 1), M (middle), and M/T (middle/tail junction). The dot labeled CTD indicates where the C-terminal domain, which is unstructured and therefore invisible, is expected to emerge from the core. (B) A schematic view of the yeast mediator illustrating the subunit composition of the head, middle, tail, and Cdk8 modules. A, from Davis, J.A. et al. (2002) Structure of the yeast RNA polymerase II holoenzyme: Mediator conformation and polymerase interaction. *Mol. Cell*, **10**, 409–415, with permission from Elsevier.

stimulate both recruitment of RNA polymerase to the closed complex and conversion of the closed complex to the open complex. Similarly, activator–Mediator interactions stimulate multiple steps in eukaryotic initiation including polymerase recruitment and steps that occur subsequent to recruitment (i.e., promoter opening and/or promoter clearance). This conclusion derives from experiments examining the consequences of deleting single Mediator subunits (Box 4.4).

In addition, to the head, middle, and tail modules, some Mediator preparations contain a fourth module termed the Cdk8 module (Figure 4.2B). The Cdk8 module may interfere with binding of the Mediator to core RNA polymerase. Not surprisingly, therefore, a number of functional studies suggest that this module actually inhibits Mediator-dependent transcriptional activation.

Box 4.4

Med23-deficient mouse cells and the mechanism of Mediator action

Many Mediator subunits are essential for Mediator integrity, and cannot be deleted without completely eliminating all Mediator function. However, certain peripheral subunits, such as Med23, can be deleted without destabilizing the remainder of the complex. Although the location of Med23 in the Mediator has not been directly determined, circumstantial evidence places it in the tail module, where it contacts a small number of activator proteins, including a factor known as Elk-1.

   Analysis of the transcription of Elk-1 target genes in Med23-deficient mouse cells indicates that activator–Mediator contacts stimulate both RNA polymerase recruitment and a step in the initiation pathway subsequent to recruitment (either promoter opening or promoter clearance) (Figure B4.4).

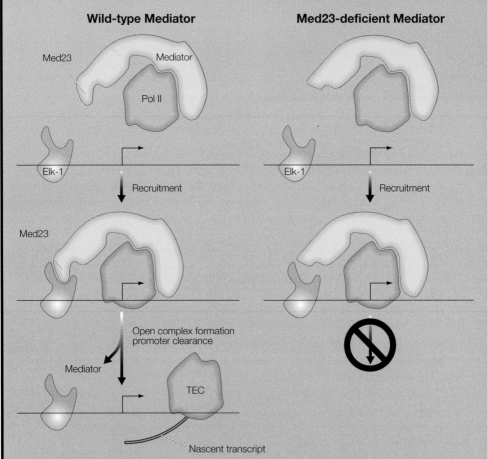

Figure B4.4 *Transcription of Elk-1 target genes in wild-type and Med23-deficient cells.* Contact between the wild-type Mediator and Elk-1 is mediated by the Med23 Mediator subunit (left). Therefore the Med23-deficient Mediator fails to contact Elk-1 (right). This reduces the efficiency of recruitment (indicated by the dotted arrow). The Elk-1–Mediator contact also stimulates open complex formation and/or promoter clearance by an unknown mechanism.

4.3.3 Release of Pol II from a paused state as a mechanism of activation

As we have seen, activators frequently recruit RNA polymerase to the promoter, and they also sometimes accelerate post-recruitment steps in initiation including open complex formation and perhaps promoter clearance. In addition, as will be discussed below, it appears that eukaryotic activators can stimulate gene expression by stimulating the elongation phase of transcription.

Release of a paused TEC by heat shock factor

Studies in *Drosophila* suggest that about 20% of class II promoters contain a paused Pol II ternary elongation complex (TEC) just downstream of the promoter (Figure 4.3). This has been best studied in the case of the heat shock genes, which encode proteins that help cells to cope with stressful conditions such as high temperature or high concentrations of agents that damage proteins (oxidants, alkylating agents, UV light, etc.). Expression of these genes is induced by an activator termed heat shock factor (HSF). In the absence of stress, HSF is monomeric and unable to bind DNA. Exposure of cells to stressful conditions induces HSF trimerization by an unknown mechanism. The trimeric form of the factor then binds to HSF recognition sites (heat shock elements) in the heat shock genes inducing their transcription.

Studies of the heat shock genes including the *hsp70* gene indicate that RNA polymerase is bound to these genes even before the cells encounter the stressful conditions that induce expression. In unstressed cells, the bound Pol II has already initiated *hsp70* transcription, but the TEC is paused just downstream of the transcriptional start site (Figure 4.3). The presence of this paused polymerase allows for an especially rapid response to a change in the environment. The binding of HSF to upstream heat shock elements in response to stress then induces the release of this polymerase from the pause site allowing completion of the transcription cycle (Figure 4.3).

Control of elongation by CTD phosphorylation

Release of Pol II from the pause site downstream of heat shock promoters depends on a factor termed positive transcription elongation factor b (P-TEFb). In the absence of P-TEFb or in the presence of the P-TEFb

By what mechanism do activator–Mediator interactions stimulate a step in transcription subsequent to Pol II recruitment?

The Mediator stimulates some step required for conversion of the preinitiation complex into a stable ternary elongation complex. But the exact step stimulated and the mechanism behind the stimulation are mysteries. An allosteric mechanism such as that discussed for activation of bacterial Pol II by λcI is a possibility. Alternatively, the Mediator may recruit an accessory factor that stimulates a postrecruitment step in the transcription cycle.

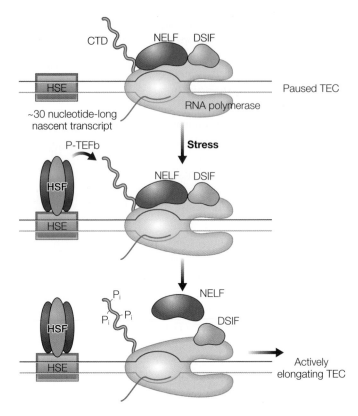

Figure 4.3 *Heat shock factor activates genes by releasing RNA polymerase from a pause site.* Prior to stress, the *Drosophila* heat shock genes contain a paused ternary elongation complex (TEC) about 30 nucleotides downstream of the transcriptional start site. Pausing at this position is induced by two core Pol II-binding factors named NELF and DSIF. Stress induces the binding of heat shock factor (HSF) to upstream heatshock elements (HSEs). This leads, by unknown mechanisms, to the recruitment of P-TEFb, a protein kinase that phosphorylates the Pol II C-terminal domain (CTD). This results in the dissociation of NELF and the resumption of transcription. DSIF remains associated with the actively elongating polymerase and is thought to increase processivity.

inhibitor 5,6-dichloro-1-β-D-ribofuranosylbenzimidazole (DRB), RNA polymerase pauses shortly after TEC formation and this pausing depends on two factors termed negative elongation factor (NELF) and DRB sensitivity inducing factor (DSIF). NELF and DSIF bind Pol II and together they may stabilize a conformation that favors pausing (Figure 4.3). Recruitment of P-TEFb to heat shock genes after exposure to stress leads to the ejection of NELF from the gene and therefore the resumption of elongation.

How does P-TEFb induce ejection of NELF? P-TEFb is a protein kinase that contains two subunits, cyclin-dependent kinase 9 (Cdk9) and its partner cyclin T. You have previously been introduced to another member of the CDK family (Chapter 3) – Cdk7, a subunit of TFIIH – which catalyzes phosphorylation of the Pol II CTD on Ser5 during promoter clearance. As we saw, CTD phosphorylation by Cdk7 is not required for transcription, but instead converts the CTD into a platform for the recruitment of enzymes that catalyze the formation of the 7-methyl-Gppp cap found at the 5' end of

> ### How does heat shock factor bring P-TEFb to the promoter?
>
> Activation of heat shock promoters by HSF appears to require recruitment of P-TEFb. But experiments to demonstrate a direct interaction between P-TEFb and HSF have yielded negative results. Therefore, we do not understand the sequence of events connecting stress-induced binding of HSF to DNA and P-TEFb recruitment.

all mRNAs. P-TEFb also phosphorylates the Pol II CTD, this phosphorylation being primarily targeted to Ser2 of the CTD heptapeptide repeats. In this case, phosphorylation appears to trigger the release of NELF from Pol II, allowing Pol II to resume elongation.

4.4 SUMMARY

Bacterial activators such as CRP and λcI stimulate transcription via direct contacts between the activators and RNA polymerase. The polymerase contains numerous surfaces that contact activators. For example, λcI contacts a surface in the σ^{70} subunit of the RNA polymerase holoenzyme, while two different activation surfaces in CRP contact two different surfaces in the α subunit of RNA polymerase. In some cases (e.g., stimulation of the lac promoter by CRP), these contacts provide added stability to the closed promoter complex, that is, they help to recruit RNA polymerase to the promoter. In other cases (e.g., stimulation of P_{RM} by λcI), these contacts stimulate promoter opening, perhaps by allosteric stabilization of the open complex relative to the closed complex.

In eukaryotes, contact between activators and core Pol II is often mediated by a 20–30-subunit protein complex known as the Mediator. The Mediator complex was discovered in biochemical fractionation experiments searching for an activity that would allow for activator-stimulated transcription in a cell-free transcription system. Many of the subunits of this complex were also discovered in a genetic screen for mutations that suppressed a growth defect due to a truncation of the Pol II CTD implying that the Mediator is required for Pol II function *in vivo*. The Mediator binds both activators and core Pol II, thereby allowing communication between the regulatory apparatus and the transcriptional machinery. These contacts are required for both Pol II recruitment and for efficient conversion of the preinitiation complex to a stable ternary elongation complex.

Activators can also regulate transcription during the elongation phase. Many genes, including heat shock genes, contain a paused Pol II TEC just downstream of the transcription start site prior to induction. In the case of the heat shock genes, binding of HSF to upstream heat shock elements stimulates release of Pol II from the pause site and completion of the transcription cycle. Pausing of Pol II depends on two inhibitory Pol II-binding factors, NELF and DSIF. HSF may stimulate the release of the paused polymerase by helping to recruit a protein kinase termed P-TEFb to the gene. P-TEFb phosphorylates the Pol II CTD, which may lead to the ejection of NELF from the paused TEC and therefore to the resumption of elongation.

PROBLEMS

1 As shown in Figure 4.2B, each basepair in DNA contains a pseudodyad axis in the plane of the basepair. Rotation of the basepair by 180° around this axis causes the two *N*-glycosidic bonds to precisely swap positions with one another. The H-bond donors and acceptors along the major groove edge of a basepair are asymmetrically arranged about the pseudodyad axis. Why is this important for sequence recognition by transcription factors? Is the pattern of H-bond donors and acceptors along the minor groove edge of a basepair symmetric or asymmetric?

2 Below are some kinetic data showing how mutations in CRP affect its ability to activate the gal promoter. Assays were carried out in the presence of wild-type CRP and in the presence of mutant forms of CRP lacking either AR1 or AR2. Explain what these data say about the relative roles of AR1 and AR2 in activation by CRP.

Activator	K_B (M^{-1})	$k_f(s^{-1})$
CRP – wild type	5×10^6	8×10^{-2}
CRP – AR1 mutant	6×10^5	7×10^{-2}
CRP – AR2 mutant	7×10^6	6×10^{-3}

3 The original mutant alleles of SRB2, SRB4, SRB5, and SRB6 were dominant, meaning they could suppress the mutant phenotypes due to a CTD truncation even in the presence of a wild-type copy of the gene. In contrast, the original mutant alleles of SRB8–SRB11 were all recessive suppressors of the CTD truncation, meaning suppression was only observed in the absence of the wild-type copy of the gene. Speculate on the nature of these dominant and recessive mutations based on the position of the corresponding subunits in the Mediator.

FURTHER READING

Sequence-specific DNA recognition

For further information about the protein DNA complexes shown in Plates 4.2, 4.3, and 4.4, the reader is referred to the online RCSB Protein Data Bank (http://www.rcsb.org). Using the PDB id numbers provided in these figures, readers can find references in which these structures were first described. In addition, the structures can be viewed and manipulated (e.g., rotated) using free molecular graphics software such as Protein Explorer (see Box 1.2).

Seeman, N.C., Rosenberg, J.M. and Rich, A. (1976) Sequence-specific recognition of double helical nucleic acids by proteins. *Proc Natl Acad Sci USA*, **73**, 804–808. *A theoretical study showing that the array of H-bond acceptors and donors present in the major groove of double-stranded DNA is sufficient to allow distinction of all possible DNA sequences.*

Bacterial activators and their targets in RNA polymerase

Guarente, L., Nye, J.S., Hochschild, A. and Ptashne, M. (1982) Mutant lambda phage repressor with a specific defect in its positive control function. *Proc Natl Acad Sci USA*, **79**, 2236–2239. *Genetic analysis of the λcI protein defining the concept of the positive control mutation.*

Li, M., Moyle, H. and Susskind, M.M. (1994) Target of the transcriptional activation function of phage lambda cI protein. *Science*, **263**, 75–77. *The identification of a suppressor allele of σ that compensates for the defect due to a pc mutation in λcI. This study is a beautiful demonstration of the use of allele-specific genetic suppression to find functionally relevant protein–protein interactions.*

Chen, Y., Ebright, Y.W. and Ebright, R.H. (1994) Identification of the target of a transcription activator protein by protein-protein photocrosslinking. *Science*, **265**, 90–92. *The use of a chemical cross-linking probe to identify the target in RNA polymerase of an activation surface in CRP.*

Niu, W., Kim, Y., Tau, G., Heyduk, T. and Ebright, R.H. (1996) Transcription activation at class II CAP-dependent promoters: two interactions between CAP and RNA polymerase. *Cell*, **87**, 1123–1134. *An analysis of the activation properties of CRP and how these properties differ depending on the target promoter.*

McClure, W.R. (1980) Rate-limiting steps in RNA chain initiation. *Proc Natl Acad Sci USA*, **77**, 5634–5638. *The abortive initiation assay and its use to define the kinetics of open complex formation.*

Eukaryotic activation domains

Hope, I.A., Mahadevan, S. and Struhl, K. (1988) Structural and functional characterization of the short acidic transcriptional activation region of yeast GCN4 protein. *Nature*, **333**, 635–640. *A deletion analysis of the GCN4 activation domain showing that activation domains are unlikely to possess well-defined globular structures.*

Radhakrishnan, I., Perez-Alvarado, G.C., Parker, D., Dyson, H.J., Montminy, M.R. and Wright, P.E. (1997) Solution structure of the KIX domain of CBP bound to the transactivation domain of CREB: a model for activator:coactivator inter-actions. *Cell*, **91**, 741–752. *A study showing that the CREB activation domain becomes ordered only upon binding to its target.*

The Mediator

Thompson, C.M., Koleske, A.J., Chao, D.M. and Young, R.A. (1993) A multi-subunit complex associated with the RNA polymerase II CTD and TATA-binding protein in yeast. *Cell*, **73**, 1361–1375. *The identification of subunits of the Mediator through a screen for suppressors of a mutation in RNA polymerase II.*

Kim, Y.J., Bjorklund, S., Li, Y., Sayre, M.H. and Kornberg, R.D. (1994) A multi-protein mediator of transcriptional activation and its interaction with the C-terminal repeat domain of RNA polymerase II. *Cell*, **77**, 599–608. *The bio-chemical identification of the Mediator.*

Asturias, F.J., Jiang, Y.W., Myers, L.C., Gustafsson, C.M. and Kornberg, R.D. (1999) Conserved structures of mediator and RNA polymerase II holoenzyme. *Science*, **283**, 985–987. *The structure of the Mediator determined by electron microscopy.*

Wang, G., Balamotis, M.A., Stevens, J.L., Yamaguchi, Y., Handa, H. and Berk, A.J. (2005) Mediator requirement for both recruitment and postrecruitment steps in transcription initiation. *Mol Cell*, **17**, 683–694. *A demonstration that the Mediator works both to stimulate Pol II recruitment and to stimulate steps in the tran-scription cycle subsequent to Pol II recruitment.*

CTD phosphorylation and the control of elongation

Lis, J.T., Mason, P., Peng, J., Price, D.H. and Werner, J. (2000) P-TEFb kinase recruitment and function at heat shock loci. *Genes Dev*, **14**, 792–803. *Demonstra-tion of the role of P-TEFb in the conversion of paused elongation complex to an active elongation complex.*

Yamaguchi, Y., Takagi, T., Wada, T., Yano, K., Furuya, A., Sugimoto, S., Hasegawa, J. and Handa, H. (1999) NELF, a multisubunit complex containing RD, cooperates with DSIF to repress RNA polymerase II elongation. *Cell*, **97**, 41–51. *A study suggesting that phosphorylation of the Pol II CTD by P-TEFb triggers the release of NELF and therefore the resumption of active elongation.*

5

Transcriptional control through the modification of chromatin structure

Key concepts

- Eukaryotic genomes are packaged into a nucleoprotein complex termed chromatin
- Chromatin exhibits multiple levels of organization and has a highly flexible structure
- Eukaryotic cells contain a multitude of coregulatory factors that catalyze covalent changes in chromatin structure
- Histone covalent modifications lead to the recruitment of regulatory factors that modulate chromatin structure thereby altering gene activity
- ATP-dependent chromatin remodeling enzymes direct non-covalent changes in chromatin structure that alter their accessibility to the transcriptional machinery

5.1 INTRODUCTION

The DNA in each eukaryotic nucleus has a contour length many orders of magnitude greater than the dimensions of the nucleus. If fully extended, the 46 chromosomes in a human cell would, for example, total about a meter in length, while a typical human nucleus is only a few microns in diameter. DNA must therefore be extensively compacted to fit inside the nucleus and to prevent the hopeless entanglement of the chromosomes, which would prevent their orderly segregation during cell division. This compaction is achieved through the packaging of the DNA into a nucleoprotein complex termed chromatin. This complex consists of roughly equal parts DNA and protein, with the major protein component being the small basic proteins termed histones.

The compaction of DNA in chromatin greatly reduces the accessibility of the DNA template to the transcriptional machinery, thereby resulting in the global repression of transcription. As we will see in this and the subsequent chapter, this global chromatin-mediated repression of transcription has allowed eukaryotic organisms to evolve a means of regulating gene expression that is not available in the other domains of life. In particular, we will see that regulatory signals direct region-specific changes in chromatin that alter the accessibility of the template to the transcriptional machinery leading to gene-specific changes in transcriptional activity.

A major way in which regulatory signals alter chromatin structure and therefore gene activity is by directing the post-translational modification, including acetylation, methylation, phosphorylation, and ubiquitylation, of histones. The combination of histone modifications associated with each gene is thought to provide the information that dictates whether or not that gene will be transcribed. The mechanisms by which these modifications regulate transcription are still being deciphered. One idea, termed the histone code hypothesis, is that these modifications serve as docking sites for regulatory factors that organize the chromatin into states that determine their competency for transcription.

The different transcriptional states that result from different combinations of histone modifications are heritable in much the same way that different alleles of a gene are heritable. Heritable phenotypes may result as much from these histone modification directed transcriptional states as from the information encoded in the DNA sequence. The realization of this fact has led to the explosive growth of a branch of molecular biology known as epigenetics, which seeks to understand the nature of these heritable transcriptional states.

As discussed previously (Chapters 1 and 4), the eukaryotic transcriptional machinery includes numerous coregulatory factors (coactivators and corepressors) that do not bind to DNA directly, but that mediate the effects of DNA-binding regulators (activators and repressors). Some of these coregulators directly transduce signals between regulators and the general transcriptional machinery – the Mediator is one coactivator that functions in this way. As we will see in Chapters 5 and 6, coactivators and corepressors also work by catalyzing changes in chromatin structure, which in turn lead to changes in the transcriptional activity of the associated genes. These chromatin-modifying coregulators are of two general types – those that catalyze non-covalent changes in chromatin structure ("chromatin remodeling") and those that catalyze the covalent modification of histones.

This chapter will begin with a brief introduction to chromatin structure. This will be followed by a discussion of post-translational histone modifications, and then by a discussion of chromatin remodeling. Finally, the chapter will conclude with an introduction to the protein motifs that recognize histone modifications. These motifs are largely responsible for reading the histone modification state and integrating the activities of the histone modifying and chromatin remodeling factors. The goals of this chapter are two-fold: the first is to evaluate the roles of covalent and non-covalent chromatin modification in gene regulation, while the second is to introduce some of the concepts required for an understanding of the epigenetic control of transcription. Chapter 6 will then elaborate on these concepts through an in-depth discussion of epigenetic transcriptional regulation.

5.2 CHROMATIN STRUCTURE

5.2.1 The nucleosome

The repeating unit of chromatin, the nucleosome, consists of a nucleosome core and linker DNA. The nucleosome core (Plate 5.1) contains a histone octamer, which includes two molecules each of histones H2A, H2B, H3, and H4 and ~150 bp of DNA. An additional ~10–50 bp of linker DNA connects the nucleosome cores and thus the average distance from the start of one nucleosome to the start of the next is about 160–200 bp. A linker histone, such as histone H1 or its close relatives, often contacts both linker DNA and the DNA associated with the core. These linker histones are frequently present in a stoichiometry of about one per nucleosome and are therefore sometimes considered a part of the nucleosome. The periodic linear array of nucleosomes along chromatin (often termed the 10 nm chromatin fiber because of the approximate 10 nm diameter of the nucleosome core) gives rise to the beads-on-a-string appearance of many chromatin preparations as viewed by electron microscopy.

The atomic resolution structure of the nucleosome core particle (Plate 5.1C, D) reveals an irregularly shaped nucleoprotein disc, about 110 Å in diameter and 80 Å in height, containing the histone octamer (Plate 5.1A) and 147 bp of DNA, which forms 1.8 left-handed superhelical turns about the octamer. All eight polypeptides in the octamer contain a so-called histone fold domain (HFD), consisting of three α-helices. The central helix in each HFD packs against the corresponding helix in another HFD thereby mediating the formation of histone dimers (termed histone fold pairs, Plate 5.1B). Histone octamer assembly begins with the formation of a histone tetramer composed of two H3:H4 dimers. The tetramer then associates with DNA followed by the association of two H2A:H2B dimers to complete the octamer (Plate 5.1A).

In addition to the HFD, each histone in the octamer contains an extended N-terminal domain; H2A and H2B also contain C-terminal domains. While these N- and C-terminal domains (termed the histone tails) are, due to disorder, mostly invisible in X-ray crystal structures, they appear to reach out from the compact globular domain of the nucleosome, enabling them to interact with other nucleosomes (see for example, the prominent H3 tail in the X-ray structure shown in Plate 5.1D).

5.2.2 Higher order chromatin structure

Analysis of chromatin in a test tube reveals that it undergoes compaction as the salt concentration in the medium is increased: this compaction is dependent upon the presence of linker histones. As the salt is increased to near physiological levels, the 10 nm fiber transitions to a 30 nm-thick fiber (Figure 5.1), the structure of which is poorly understood. It may result from folding of the 10 nm fiber into a simple, smooth helix containing about six nucleosomes per turn. However, this model is controversial – and other more complex models for the structure of the 30 nm fiber are gaining in popularity.

Compaction of the 10 nm fiber to form the 30 nm fiber depends upon the N-terminal tails and is promoted by linker histones. Relatively normal 10 nm fibers can be assembled in a test tube using mutant histones lacking the N-terminal tails, but these fail to form 30 nm fibers in the presence of linker histones and high salt. This supports the idea that high order chromatin folding involves contacts between the N-terminal tails and nucleosome cores.

Inside nuclei, the 30 nm chromatin fiber undergoes further compaction (Figure 5.1). For example, electron and fluorescence microscopy reveal that chromatin inside nuclei is predominantly in the form of ~100 nm-thick fibers. While the structure of these so-called "chromonema fibers" is not understood, a speculative model suggests that they result from the folding of the 30 nm fiber back on itself. Chromonema fibers undergo further compaction perhaps by the formation of loops anchored to a protein scaffold.

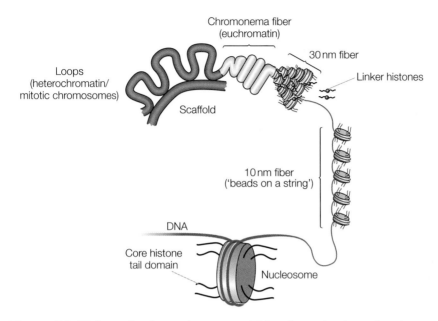

Figure 5.1 *Higher order chromatin structure.* This schematic view of a piece of chromatin illustrates the various levels of organization of chromatin. In the first level of organization, naked DNA wraps around a histone octamer to form nucleosomes. An array of nucleosomes along a length of DNA can be visualized in the electron microscope as the 10 nm chromatin fiber. In the presence of linker histones, the 10 nm chromatin fiber further condenses to form the 30 nm fiber. This is thought to involve the folding of the 10 nm fiber into a helix, the exact structure of which is not known. The 30 nm fiber is stabilized by interactions between histone tails and the globular domains of adjacent nucleosomes. Inside cells, 30 nm chromatin fibers further condense to form chromonema fibers. The nature of the transition is not understood – one possibility is that it involves the folding back of the 30 nm fiber upon itself as illustrated here. In interphase nuclei, most of the chromatin (the euchromatin) is in the form of chromonema fibers. Some of the chromatin (the heterochromatin) is further condensed, perhaps by the formation of loops anchored to a protein scaffold. In mitotic cells, all the chromatin assumes this highly condensed conformation. Adapted from Horn, P.J. and Peterson, C.L. (2002) Molecular biology. Chromatin higher order folding–wrapping up transcription. *Science*, **297**, 1824–1827.

5.2.3 Euchromatin and heterochromatin

During mitosis, the chromatin becomes highly condensed leading to the formation of mitotic chromosomes that can be easily visualized in dividing cells by light microscopy. Once mitosis is complete, the chromosomes decondense and become invisible in the light microsope, which is how

Mitosis – the process of nuclear and cellular division producing two identical daughter cells each containing a complete set of chromosomes. During mitosis, the nuclear membranes dissolve, the chromosomes condense, and one copy of each chromosome is pulled to each pole by the mitotic apparatus. The nuclear membranes reform, and the cell pinches in two between the two nuclei.

they remain throughout interphase (the period between the mitotic cycles). In the 1920s, Emil Heitz noticed that a small fraction of the chromatin failed to decondense during interphase and he referred to this type of chromatin as heterochromatin ("different chromatin") to distinguish it from euchromatin ("true chromatin"), which does decondense.

Subsequent studies have shown that euchromatin is composed of ~100 nm-thick fibers (i.e., the chromonema fibers discussed in the previous section), while heterochromatin is a further condensed form of these ~100 nm fibers. Furthermore, while euchromatin is diffusely localized throughout the nucleoplasm, much of the heterochromatin is found adjacent to the nuclear membrane (Figure 5.2). A final distinguishing characteristic of heterochromatin is its tendency to replicate significantly later than euchromatin during each S phase.

S phase – the period in the cell cycle when the chromosomes replicate.

By carefully following chromosomes through the mitotic cycle as they condensed and decondensed, Heitz showed that much of the heterochromatin is associated with the

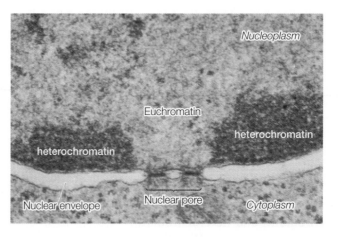

Figure 5.2 *Transmission electron micrograph of part of a nucleus.* The heterochromatin is the densely staining material lining the nuclear envelope, while the euchromatin is the thread-like material found throughout the nucleoplasm. Note the absence of heterochromatin from the region around the nuclear pore. From Fawcett, D.W. (1981) *The Cell.* W.B. Saunders, Philadelphia, p. 271, with permission from Elsevier.

centromeres (chromosomal regions to which the mitotic spindles attach) and the telomeres (ends of the chromosomes). Very few genes map to these regions, leading Heitz to conclude that "euchromatin is genicly active, heterochromatin genicly passive". Subsequent studies have largely confirmed this view – the condensed conformation of heterochromatin is thought to be incompatible with the transcription of most genes. Furthermore, transcriptional activation domains are able to induce the decondensation of heterochromatin into euchromatin and this probably involves the recruitment of coactivators that catalyze covalent and non-covalent changes in chromatin structure.

Since centromeres and telomeres are almost always heterochromatic in normal nuclei, the heterochromatin associated with these regions is termed constitutive heterochromatin. This form of heterochromatin has critical roles in chromosome maintenance (the accurate replication of chromosomes and the faithful distribution of newly synthesized chromosomes to daughter cells during mitosis), but does not normally have a role in gene silencing since these regions are mostly devoid of genes. However, a second form of heterochromatin, termed facultative heterochromatin, forms at discrete regions within the primarily euchromatic chromosome arms for the express purpose of silencing gene expression. As will be discussed in Chapter 6, facultative heterochromatin plays major roles in the epigenetic control of transcription.

5.3 HISTONE MODIFICATION

Histones are subject to many post-translational covalent modifications, including lysine acetylation, lysine and arginine methylation, serine phosphorylation, and lysine ubiquity-lation (Figures 5.3, 5.4). While most of these modifications map to the N-terminal tails of the histones, a few map to the central and C-terminal domains. Histone modifications appear to play a pivotal role in the determination of transcriptional state.

The means by which these modifications regulate chromatin function in general and transcription in particular is the subject of intensive research. In general, there are two possibilities, which are not mutually exclusive of one another:

1 These modifications could directly alter chromatin folding. As noted above, the

> **How do covalent histone modifications directly modulate chromatin structure?**
>
> One possibility, suggested by X-ray crystallography, is that positively charged lysine side chains in histone tails are attracted to negatively charged surfaces in other nucleosomes leading to chromatin condensation. Acetylation, which neutralizes the lysine side chains, might therefore be expected to lead to chromatin decondensation. While this is a nice theory, experimental proof is lacking.

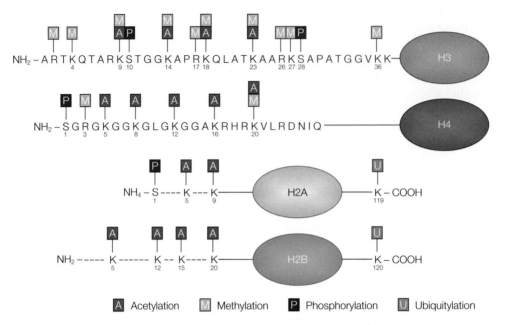

Figure 5.3 *The post-translational modifications of the four core histones.* The ellipses represent the globular core domains. Most of the known modifications of the N- and C-terminal tails are shown.

histone N-terminal tails are essential for the higher order folding of chromatin. Thus, covalent modification of these tails could modulate higher order chromatin structure. Although it has been difficult to prove that this occurs inside living cells, experiments on chromatin assembled *in vitro* provide support for this idea. In particular, acetylation of histone N-terminal tails can lead to decondensation of chromatin.

2 Modified residues in histones could alter the ability of histones to recruit non-histone proteins to chromatin, which could, in turn, alter the ability of the transcriptional machinery to recognize the template. As will be discussed extensively in this and the following chapter, a growing body of evidence supports this idea. For example, many factors that regulate gene expression contain bromodomains, which bind to acetyl-lysine residues, or chromodomains, which bind to methyl-lysine residues.

The modification state of histones has been proposed to constitute a combinatorial code that sets the transcriptional state of each gene. While the concept of a histone modification code is useful, we will see that it is likely an oversimplification to assume that each combination of histone post-translational modifications is associated with a unique transcriptional state. Rather the transcriptional state of a gene is probably a product not only of the combination of histone modifications associated with that gene, but also of the developmental history of the tissue. This is because developmental history determines what array of transacting factors (e.g., bromodomain-containing factors, chromodomain-containing factors, etc.) are available to interact with the modified histones.

5.3.1 Lysine acetylation: diverse roles in gene activation

The first histone modification to be studied in detail was lysine acetylation. In this modification, enzymes termed histone acetyltransferases (HATs) transfer an acetyl group from acetyl-coenzyme A to the ε-amino group of a lysine residue (Figure 5.4A). This converts the amino group, which is positively charged at physiological pH, to a neutral amide. Acetylation is readily reversible: in addition to multiple HATs, eukaryotic cells also contain multiple histone deacetylases (HDACs), which catalyze the release of acetate regenerating the free lysine side chain.

The connection between histone acetylation and gene activation

Observations dating back to the early 1960s suggested a connection between histone acetylation and the active transcriptional state. For example, in 1964 V.G. Allfrey and coworkers showed that acetylation of histones, while not interfering with the ability of the histones to complex with DNA, markedly reduced their ability to inhibit transcription by partially purified preparations of calf thymus RNA polymerase. Based on this result, it was hypothesized that histone acetylation might provide a reversible means of regulating gene activity.

In the late 1980s, a solid connection between gene activity *in vivo* and histone acetylation was made through the development and application of the chromatin immunoprecipitation (ChIP) assay (Box 5.1). In these experiments, chromatin was isolated from rabbit reticulocytes and fragmented into small pieces (typically a few nucleosomes in length). Immunoprecipitation with antibodies against *N*-acetyl-lysine was then used to separate hyperacetylated chromatin fragments (which bind the antibody) from hypoacetylated chromatin fragments (which fail to bind the antibody). Analysis of the two fractions showed that the hyperacetylated fraction was highly enriched for transcriptionally active genes such as the globin genes, while the hypoacetylated fraction was enriched for transcriptionally silent genes, such as the ovalbumin gene.

Although the ChIP results were suggestive, they did not prove that histone acetylation was a prerequisite for gene activation. For example, it was possible that levels of histone acetylation changed only as a consequence of activation. Acceptance of Allfrey's idea, proposed 30 years earlier, that histone acetylation has a direct role in the regulation of gene expression came in the mid-1990s as a result of efforts to purify and

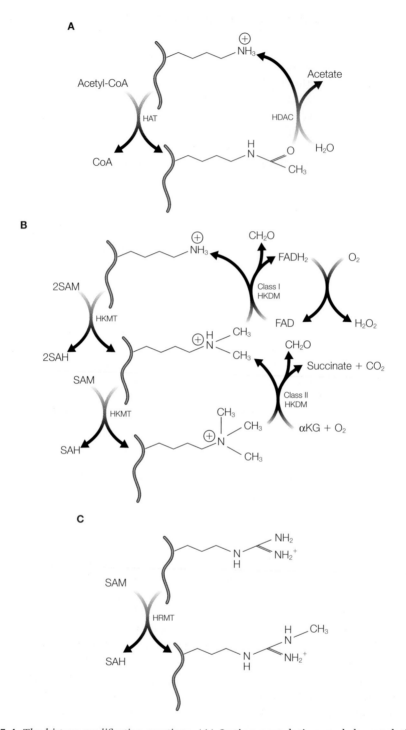

Figure 5.4 *The histone modification reactions.* (A) Lysine acetylation and deacetylation. (B) Lysine methylation and demethylation. (C) Arginine methylation. A single methylation reaction leading to monomethyl-arginine is shown. Monomethyl-arginine can be methylated a second time on the same amino group to produce the asymmetric dimethyl species or on the other primary amino group to produce the symmetric dimethyl species.

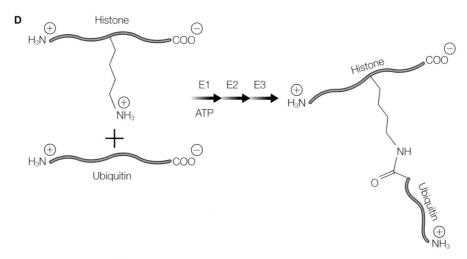

Figure 5.4 (*continued*) *The histone modification reactions.* (D) Lysine ubiquitylation. Ubiquitin is a 76 amino acid protein. Attachment via its carboxyl terminus to a histone lysine side chain is catalyzed by three sequentially acting enzymes, E1, E2, and E3. Each cycle of ubiquitylation consumes a single ATP and produces AMP and pyrophosphate. CoA, coenzyme A; FAD and $FADH_2$, oxidized and reduced forms of flavin adenine dinucleotide; HAT, histone acetyltransferase; HDAC, histone deacetylase; HKDM, histone lysine demethylase; HKMT, histone lysine methyltransferase; HRMT, histone arginine methyltransferase; α-KG, α-ketoglutarate; SAH, *S*-adenosylhomocysteine; SAM, *S*-adenosylmethionine.

characterize the enzymes that catalyze histone acetylation and deacetylation. One such set of experiments led to the discovery that the yeast gene Gcn5 encoded a nuclear HAT.

Gcn5 had been previously identified in several genetic screens that suggested a role for its product in transcriptional activation. One such screen turned up five genes termed ADA1 through ADA5, which were required for the functioning of a number of activation domains including the VP16 activation domain in yeast cells. Ada4 was found to correspond to Gcn5 and, as we now know, the other Ada proteins are components along with Gcn5 of a multisubunit HAT complex called the SAGA complex. The finding that a protein required for promoter selective gene activation was the catalytic subunit of a HAT cemented the idea that histone acetylation played a direct role in gene activation. Further confirmation of this conclusion comes from the discovery that genes required for transcriptional repression (e.g., RPD3, SIN3) encode subunits of histone deacetylase complexes.

HAT and HDAC complexes

As implied above, HATs are often found in large protein complexes (Table 5.1). For example, Gcn5 is a component of the yeast SAGA complex; and human cells contain multiple complexes with extensive

Box 5.1

The chromatin immunoprecipitation assay

The chromatin immunoprecipitation (ChIP) assay has become ubiquitous in the transcription literature. In essence, it is a technique that allows one to determine where a protein of interest is bound within a genome of interest. A ChIP assay to determine where protein X is bound to the human genome involves the following steps (Figure B5.1):

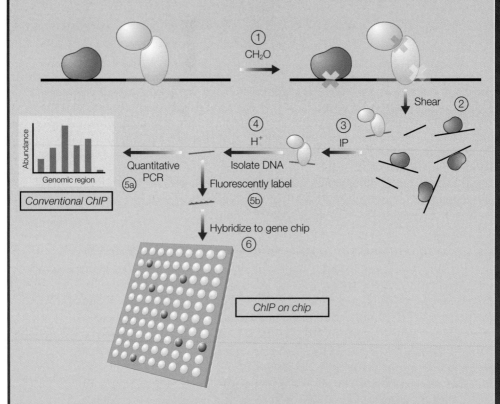

Figure B5.1 *The chromatin immunoprecipitation assay.* See text for an explanation of steps 1–6. The bar graph at bottom left represents the results of a conventional ChIP assay in which selected regions of the genome are amplified with pairs of PCR primers. The abundance of the resulting PCR product indicates the frequency with which protein X is found in association with a particular region. The grid at bottom right represents a gene chip onto which segments of DNA representing a large fraction of a genome have been arrayed. A real chip of this kind might contain tens of thousands of arrayed DNA fragments. The fluorescently labeled immunoprecipitated DNA is hybridized to this chip to determine the regions of the genome with which protein X associates.

1 Cultured human cells are treated with a cross-linking agent (usually formalde-hyde) to covalently cross-link proteins to DNA as well as proteins to proteins. The goal is to preserve interactions that exist in the cell prior to cross-linking, not to create new interactions.

2 Chromatin is extracted from the cells and broken into small fragments, the size of which is on the order of one to two nucleosomes. Fragmentation is achieved by sonication and/or treatment with non-specific nucleases.

3 The mixture of chromatin fragments is subjected to immunoprecipitation with anti-bodies against protein X (or a particular covalent modification in protein X). As a result, chromatin fragments containing protein X (or a particular modified form of protein X) are precipitated, while other chromatin fragments remain soluble.

4 The formaldehyde cross-links in the precipitate are reversed by low pH treatment, the proteins are extracted, and the protein-free DNA fragments are isolated.

5a To determine whether or not any particular genomic region was immuno-precipitated (indicating that it was bound by protein X), the precipitated DNA fragments are used as the template in a quantitative polymerase chain reac-tion (PCR) using PCR primers that amplify regions of interest.

5b, 6 A variation on the ChIP assay is the so-called ChIP on chip assay. In this technique, the DNA from the immunoprecipitate is fluorescently labeled, and then used to probe a DNA microarray (a gene chip, hence the name "ChIP on chip") on which has been arrayed DNA fragments representing a large sample of the genome. In this way, one can determine the genome-wide distribution of a protein in a single experiment.

ChIP assays can be used to determine where non-histone proteins (e.g., HP1) are bound to chromatin. They can also be used to determine the positions of vari-ous histone post-translational modifications in the genome by using antibodies that only recognize specifically modified forms of histones (e.g., antibodies that only recognize the trimethyl-lysine 9 form of histone H3).

similarity to the yeast SAGA complex. A second family of HAT complexes, the MYST complexes, contains catalytic subunits related to the yeast Esa1 protein. Different HAT complexes have different substrate specificities.

HDACs are also components of multisubunit complexes. Rpd3, which is the best characterized of the yeast HDACs, is found in two multisub-unit complexes termed the Sin3 complex and the NuRD (nucleosome remodeling and deacetylation) complex.

Targeted versus global histone acetylation and deacetylation

The finding that histone acetylation and deacetylation are required for activation and repression can be interpreted in multiple ways. For example,

Table 5.1
Histone acetyltransferase (HAT) complexes

Complex	Catalytic subunit[1]	Substrate specificity	Tra1-related subunit
GNAT family complexes[2]			
Yeast SAGA	yGcn5 (BrD)	H3/H2B	Tra1
Human STAGA	hGcn5 (BrD)	H3/H2B	TRRAP
Human PCAF	PCAF (BrD)	H3/H4	PAF400
MYST family complexes[3]			
Yeast NuA4	Esa1 (ChD)	H4/H2A	Tra1
Human Tip60	Tip60 (ChD)	H4/H2A	TRRAP
Fly MSL	MOF (ChD)	H4 K16	?
Yeast SAS	Sas2 (ChD)	H4 K16	?

[1]In addition to a HAT domain, each catalytic subunit contains either a bromodomain (BrD) or a chromodomain (ChD) as indicated.
[2]All the GNAT (for Gcn5-related acetyltransferase) family complexes contain a catalytic subunit with homology to yeast GCN5.
[3]All the MYST (for MOF, Ybf2, Sas2, Tip60) family complexes contain a catalytic subunit with homology to yeast Esa1.

it is possible that promoter-specific transcription factors recruit and therefore target HATs and HDACs to specific genes leading to a change in their acetylation state and therefore to a change in transcriptional state. It is also possible that global histone acetylation is necessary to set up a state that is permissive for gene activation and that the gene-specific events leading to changes in transcriptional state do not involve targeted changes in acetylation. While these two mechanisms are not mutually exclusive and may both operate, it is clear that targeted histone acetylation and deacetylation is an important part of gene regulation.

Studies of the Myc family of bHLH factors (see Chapter 4 for a description of the bHLH domain) have provided evidence for targeted histone acetylation and deacetylation in gene regulation (Figure 5.5). This family of factors, which includes Myc, Mad, and Max, controls the switch between cell proliferation and cell differentiation. Mutations in the genes encoding these factors result in misregulation of this switch and are associated with a large fraction of human cancers.

The Myc:Max heterodimer functions as a transcriptional activator to stimulate the expression of genes required for cell proliferation, while the Mad:Max heterodimer functions as a repressor to shut off these same targets and thereby allow cells to differentiate. An activation domain in Myc binds to TRRAP, which is the single subunit shared by both SAGA and MYST family HAT complexes (Table 5.1). The TRRAP–Myc

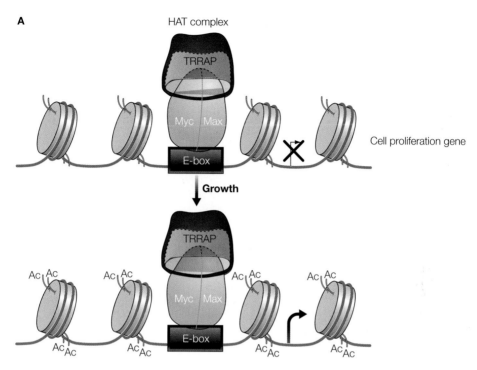

Figure 5.5 *Targeted histone acetylation and deacetylation by bHLH family transcription factors.* (A) Myc functions as a heterodimer with Max to activate genes required for cell proliferation. Under conditions that favor growth, the Myc:Max heterodimer binds to E-boxes in cell proliferation genes and recruits histone deacetyltransferase (HAT) complexes by binding to TRRAP, a polypeptide common to a number of mammalian HAT complexes including the STAGA complex and the Tip60 complex. The HAT then acetylates nearby nucleosomes leading to gene activation.

interaction leads to the recruitment of HAT complexes to Myc targets genes and the targeted acetylation and activation of these genes (Figure 5.5A). While Myc binds TRRAP, Mad binds Sin3, a subunit of the Sin3–HDAC complex. Thus, the Mad:Max heterodimer recruits the Sin3–HDAC complex leading to targeted histone deacetylation and repression (Figure 5.5B).

Context dependence of histone acetyl-lysine function

In general, high levels of histone acetylation correlate with gene activation, while low levels of acetylation correlate with repression. However, this simple statement masks the true complexity of the situation. Through the use of ChIP on chip experiments (see Box 5.1), it has been possible to systematically determine the acetylation levels for a number of histone lysine residues throughout the entire yeast genome. These studies show that acetylation of several lysines in histone H3 correlate with high

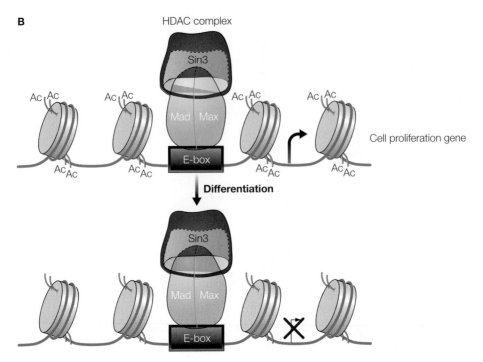

Figure 5.5 (*continued*) *Targeted histone acetylation and deacetylation by bHLH family transcription factors.* (B) Mad functions as a heterodimer with Max to repress genes required for cell proliferation. Under conditions that favor differentiation, the Mad:Max heterodimer binds to E-boxes in cell proliferation genes and recruits the Sin3–histone deacetylase (HDAC) complex by binding to Sin3. The HDAC complex then deacetylates nearby nucleosomes leading to repression.

levels of gene activity, while acetylation of histone H4 lysine 16 correlates with low levels of gene activity.

The correlation between H4 lysine 16 acetylation and repression is not universal, however. There are a number of known connections between acetylation of this lysine and activation, one of which has been revealed by studies of dosage compensation in *Drosophila*. In fruit flies, males contain one X chromosome, while females contain two. To compensate for this difference in X chromosome dosage between the sexes, males upregulate transcription of X-linked genes by a factor of two. This upregulation results from the acetylation of histone H4 lysine 16 along much of the X chromosome by MSL, a MYST family HAT complex.

Thus, acetylation of histone H4 lysine 16 sometimes correlates with gene activation and other times with gene repression. This illustrates a general theme in eukaryotic gene regulation: there is no universal link between the modification of a particular histone residue and gene activity. The effect of any given modified residue depends on the context

Table 5.2

Histone methyl marks

Residue	Enzymes[1]	Function
H3 Lys9	Su(var)3-9 [Dm]	Repression
	SUV39H1 [Hs]	
	Clr4 [Sp]	
H3 Lys27	E(z) [Dm]	Repression
	EZH2 [Hs]	
H4 Lys20	SUV4-20H [Hs]	Repression
H3 Lys4	Trithorax [Dm]	Activation
	Ash1 [Dm]	
	Mll [Hs]	
	Set1 [Sc]	
H3 Lys36	Set2 [Sc]	Activation
H3 Lys79	Dot1 [Sc]	Activation
H3 Arg17	Carm1 [Hs]	Activation
H4 Arg3	PRMT1 [Hs]	Activation

[1]The species in which each enzyme is found is given in square brackets:
Dm, *Drosophila melanogaster*; Hs, *Homo sapiens*; Sc, *Saccharomyces cerevisiae*;
Sp, *Schizosaccharomyces pombe*.

of that modification relative to other modified residues, as well as on the cellular environment. Clearly the histone modification code is a highly sophisticated code that cannot be deciphered in a straightforward way.

5.3.2 Lysine methylation: a chemically stable histone mark

Although histone methylation was discovered at about the same time as histone acetylation, the intensive analysis of histone methylation as an important aspect of transcriptional control did not begin until much more recently. Methyl-lysine residues in histone N-terminal tails can function as either activating or repressing marks depending on the context of the modified lysine residue (Table 5.2). As will be discussed extensively in Chapter 6, histone methylation plays central roles in the epigenetic activation and silencing of gene expression.

Histone lysine methyltransferases

Just as cells contain multiple HATs, they also contain multiple histone lysine methyltransferases (HKMTs). These enzymes transfer a methyl group from *S*-adenosylmethionine to the lysine ε-amino group, which can be methylated multiple times to produce mono-, di-, and trimethyl-lysine (Figure 5.4B). Many HKMTs contain a catalytic domain termed a SET domain, SET being an acronym for the three *Drosophila* proteins, Su(var)3-9, Enhancer of zeste, and Trithorax, that first defined the family. Long before their roles as HKMTs were known, these three proteins were identified through studies of transcriptional regulation during fruit fly development, thus demonstrating the central role of histone methylation in this process (see Chapter 6). SET domain HKMTs are highly substrate specific: most will methylate only a single histone lysine residue (Table 5.2).

Histone lysine demethylases

In contrast to lysine acetylation, lysine methylation cannot be readily reversed by hydrolysis. As will be discussed in Chapter 6, histone methylation plays a central role in the formation of transcriptional states that are stable enough to be passed on from cell generation to cell generation, a fact that may be related to the chemical stability of methyl-lysine. Despite the stability of methyl-lysine, enzymatic histone methyl-lysine demethylation does occur. The histone lysine demethylases (HKDMs) contend with the inherent stability of methyl-lysine by employing oxidative mechanisms to break the carbon–nitrogen bond, releasing the carbon as formaldehyde. These enzymes fall into two classes depending on the nature of the oxidant and the catalytic mechanism (Figure 5.4B).

5.3.3 Cross talk between histone marks

Cross talk between different types of histone post-translational modifications, in which one modification positively or negatively influences the extent of another modification, plays an important role in determining the ultimate pattern of histone modification throughout the genome. Three types of histone modification that are often involved in this type of cross talk are arginine methylation, serine phosphorylation, and lysine ubiquitylation.

What is the molecular basis for cross talk between histone modifications?

One unproven possibility is that one modification forms a docking site for enzymes that catalyze subsequent modifications.

Cross talk between arginine methylation, serine phosphorylation, and lysine acetylation

Mono- or dimethylation of histone arginine residues often correlates with gene activation. For example, activation of the β-globin locus in chicken erythroid cells correlates with the methylation of histone H4 arginine

3 by protein arginine methyltransferase 1 (PRMT1). PRMT1-mediated methylation stimulates the acetylation of histones H3 and H4 leading to activation (Figure 5.6A). The effect of arginine methylation on acetylation is thought to be direct since *in vitro* methylation of nucleosomes with PRMT1 stimulates *in vitro* acetylation by HATs present in erythroid cell nuclear extracts.

Histone phosphorylation has also been positively correlated with histone acetylation in a number of instances. For example, activation of certain yeast genes (e.g., the INO1 gene) requires the sequential phosphorylation of histone H3 serine 10 and acetylation of histone H3 lysine 14 (Figure 5.6B).

Cross talk between lysine ubiquitylation and lysine methylation

Ubiquitin is a 76 amino acid protein that becomes conjugated to a wide variety of target proteins via an amide linkage between its C-terminal carboxyl group and the ε-amino group in a lysine residue of the target protein. The formation of these branched proteins requires the sequential action of three enzymes (E1, E2, and E3; Figure 5.4D). Eukaryotic cells contain dozens of E3 family enzymes (sometimes termed ubiquitin ligases) to ensure the specificity of ubiquitin conjugation. Within the nucleosome core particle, the H2A and H2B C-terminal domains are both targets for ubiquitylation.

In a particularly intricate example of cross talk between histone modifications, the activator Gal4 triggers the ubiquitylation of H2B at lysine 123 and this leads to the methylation of histone H3 at lysines 4 and 36 (Figure 5.6C), a combination of modifications associated with the active transcriptional state.

5.4 ATP-DEPENDENT CHROMATIN REMODELING ENZYMES

5.4.1 A diverse family of chromatin remodeling complexes

While the previous section focused on the role of histone covalent modification in the regulation of gene activity, non-covalent changes in chromatin structure, largely brought about by ATP-dependent chromatin remodelers, also play critical roles in the regulation of transcriptional state. Like HATs and HDACs, the remodelers were first identified through genetic approaches. Genetic screens in yeast led to the identification of two sets of genes, the SWI (switch) genes, which are required for mating-type switching, and the SNF (sucrose non-fermenting) genes, which are required for sucrose fermentation. The SWI/SNF genes were found to encode the subunits of an 11 polypeptide coactivator complex. The Snf2 subunit of this complex is an ATPase, the activity of which is greatly

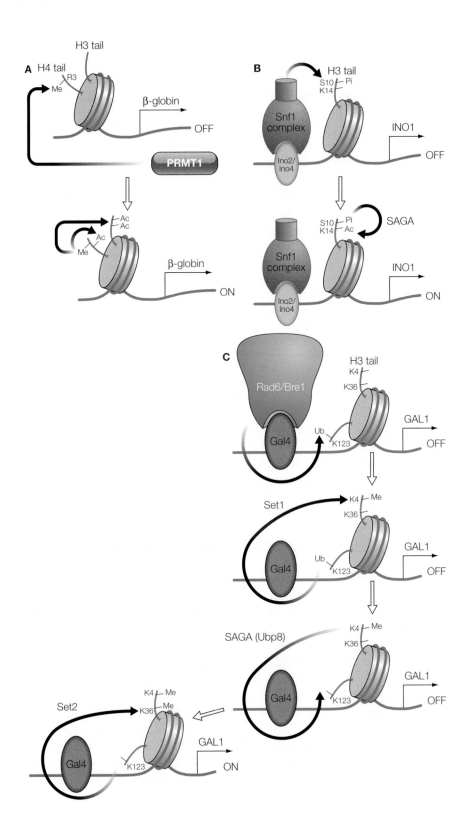

stimulated by chromatin. As we will see, Snf2 and other related ATPases couple the energetically favorable hydrolysis of ATP to nucleosome movement or changes in nucleosome structure.

Since the discovery of the SWI/SNF complex, numerous additional chromatin remodeling complexes have been identified in many eukaryotic organisms. All of these complexes contain a catalytic subunit with homology to Snf2 (Table 5.3). Major subfamilies of chromatin remodeling complexes include the SWI/SNF subfamily, the imitation switch (ISWI) subfamily, and the Mi-2/CHD subfamily. In addition to roles in transcription, some of these complexes are required for other aspects of DNA metabolism such as the assembly of nucleosomes onto newly replicated DNA.

5.4.2 ATP-fueled motors for increasing DNA accessibility

Snf2 and its relatives are molecular motors that use the energy of ATP hydrolysis to move along the DNA. By remaining anchored to the histone octamer while tracking on DNA, these enzymes serve to push or pull nucleosomes along the template. More specifically, they are proposed to stimulate the dissociation of a stretch of DNA from the histone octamer leading to the formation of a DNA bulge. ATP-fueled movement of the ATPase along the DNA then advances the DNA along the octamer (Figure 5.7A).

Different remodeling enzymes induce the dissociation of different size segments of DNA: ISWI family complexes induce the dissociation of small segments (~10 bp), while SWI/SNF family complexes induce the dissociation of larger segments (>50 bp). As a result, ISWI complexes are restricted to generating movement of histone octamers along DNA, while SWI/SNF complexes are able to generate more dramatic changes in chromatin structure. SWI/SNF complexes are, for example, able to induce the dissociation of H2A:H2B dimers from nucleosomes (Figure 5.7B).

Figure 5.6 (*opposite page*) *Cross talk between histone modifications.*
(A) Methylation of histone H4 Arg3 at the mammalian β-globin promoter leads to acetylation of the H3 and H4 tails at multiple positions resulting in gene activation. HAT, histone acetyltransferase; PRMT1, protein arginine methyltransferase 1. (B) The Ino2/Ino4 activator recruits a complex containing the Snf1 protein kinase to the INO1 promoter. Snf1 phosphorylates histone H3 Ser10. This leads to acetylation of histone H3 Lys 14 by the SAGA complex, resulting in gene activation. (C) Gal4 recruits a complex of Rad6 (a ubiquitin E2) and Bre1 (a ubiquitin E3) to the GAL1 promoter. This leads to ubiquitylation of histone H2B Lys123, which in turn leads to methylation of histone H3 Lys4 by Set1. Lys4 methylation triggers deubiquitylation of H2B Lys123 by Ubp8, a subunit of the SAGA complex. This, in turn, leads to methylation of histone H3 Lys36 by Set2 and thus to gene activation.

Table 5.3

ATP-dependent chromatin remodeling complexes

Subfamily	Examples (species; catalytic subunit)	Catalytic subunit domains
SWI/SNF	SWI/SNF (Sc; Snf2) RSC (Sc; Sth1) Brm (Dm; Brm) hBrm (Hs; hBrm)	ATPase + bromodomain
ISWI	ISW1 (Sc; Iswi) NURF (Dm; ISWI) ACF (Dm; ISWI) hACF (Hs; hSNF2h)	ATPase + SANT[1]
CHD1	dMi-2 (Dm; dMi-2) NURD (Hs; Mi-2α)	ATPase + chromodomain

ACF, ATP-utilizing chromatin assembly and remodeling factor; Brm, Brahma; Dm, *Drosophila melanogaster*; Hs, *Homo sapiens*; NuRD, nucleosome remodeling and deacetylation (complex); NURF, nucleosome remodeling factor; RSC, remodels the structure of chromatin; SANT, Swi3, Ada2, N-CoR, TFIIIB domain; Sc, *Saccharomyces cerevisiae*.
[1]The SANT domain is a motif found in a wide variety of regulatory factors and can mediate both protein–protein and protein–DNA interactions. It helps to decode the histone modification code by binding to unmodified histone tails.

H2A:H2B dimer removal can lead to multiple alternative consequences, including the following:

1 H2A:H2B dimer removal may sometimes lead to complete disassembly of nucleosomes, generating a stretch of naked DNA.
2 The missing H2A:H2B dimers can sometimes be replaced by a variant histone dimer containing one of several minor H2A isoforms such as H2A.Z. Minor histone isoforms have multiple functions in DNA metabolism, and may mark genes for activation or silencing in ways that are not well understood.

The ability of chromatin remodeling enzymes to mobilize nucleosomes, alter nucleosome structure, or completely remove nucleosomes enables them to increase the accessibility of DNA templates to the transcriptional machinery. This is important because the assembly of nucleosomes onto promoter DNA inhibits recognition of the promoter by the components of the preinitiation complex. ATP-dependent remodeling complexes are thought to alleviate this inhibition by moving histone octamers out of the way. In addition to being required for preinitiation complex assembly, chromatin remodeling complexes may be required for elongation since intact

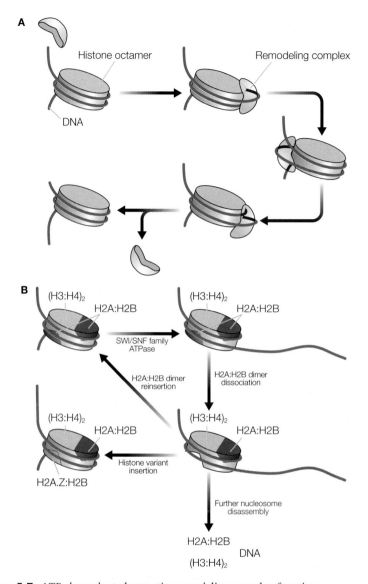

A

Histone octamer   Remodeling complex

DNA

B

(H3:H4)₂
H2A:H2B

SWI/SNF family
ATPase

(H3:H4)₂
H2A:H2B

H2A:H2B dimer
reinsertion

H2A:H2B dimer
dissociation

(H3:H4)₂
H2A:H2B

(H3:H4)₂
H2A:H2B

H2A.Z:H2B

Histone variant
insertion

Further nucleosome
disassembly

H2A:H2B
                    DNA
(H3:H4)₂

Figure 5.7 *ATP-dependent chromatin remodeling complex function.*
(A) Chromatin remodeling complexes are thought to induce the
dissociation of a stretch of DNA from the nucleosome and then track
along the DNA pulling the DNA bulge along the histone octamer surface.
As a result, the DNA advances along the nucleosome. (B) If a chromatin
remodeling complex (such as SWI/SNF) induces the dissociation of a large
enough stretch of DNA from the edge of the nucleosome, this can result
in the almost complete loss of contact between DNA and one of the
H2A:H2B dimers. As a result, the dimer can dissociate from the octamer.
This dissociated dimer can be replaced by another H2A:H2B dimer.
Alternatively, the missing dimer can be replaced by a histone variant
dimer containing H2A.Z, an H2A variant, or some other variant dimer.
Lastly, the nucleosome destabilized by H2A:H2B dissociation can further
disassemble to produce naked DNA.

nucleosomes severely hamper the progress of the ternary elongation complex. By partially or fully disassembling nucleosomes within the transcription unit, chromatin remodeling complexes thus facilitate elongation.

While the ability of these factors to render templates accessible to the transcriptional machinery seems to be most compatible with a role in transcriptional activation, ATP-dependent chromatin remodeling complexes also have roles in repression. The NuRD complex, a Mi-2/CHD family complex, contains histone deacetylase subunits in addition to the Mi-2 ATP-dependent chromatin remodeling subunit. Perhaps by altering chromatin structure, Mi-2 renders the histones accessible to the HDAC subunits of the complex. The resulting deacetylation of the chromatin might then result in transcriptional silencing.

5.4.3 Targeting of chromatin remodeling complexes

While chromatin remodeling complexes may sometimes act in a global way to establish a state that is permissive for transcription, ample evidence indicates that these complexes also act in a gene-specific manner. At least three mechanisms have been identified that target ATP-dependent chromatin remodeling complexes to specific genes:

1 Many sequence-specific factors (e.g., Gal4, Myc, and glucocorticoid receptor) bind chromatin remodeling factors and recruit them to DNA in a gene-specific manner. The SWI/SNF complex contains multiple surfaces that bind to acidic activation domains.
2 The SWI/SNF complex also binds directly to RNA polymerase II, providing another means of targeting this chromatin remodeling complex to active genes. The SWI/SNF complex or other chromatin remodeling complexes may move down the template with Pol II eliminating or partially disassembling nucleosomes in advance of the ternary elongation complex, thereby allowing elongation to continue.
3 Chromatin remodeling complexes contain domains that target them to specifically modified forms of histones. For example, Snf2 and its close relatives contain bromodomains in addition to ATPase domains. As will be discussed in the following section, bromodomains bind histone acetyl-lysine residues. Thus, gene-specific acetylation of histones by HAT complexes leads to the gene-specific recruitment of chromatin remodeling complexes.

5.5 PROTEIN MOTIFS THAT RECOGNIZE MODIFIED HISTONES

As discussed earlier, histone post-translational modifications direct changes in transcriptional state. Interpretation of these modifications relies on protein domains that recognize and bind to modified histones in a context-dependent manner. The following discussion will focus on two such motifs, the chromodomain, which binds di- and trimethyl-lysine residues, and the bromodomain, which binds acetyl-lysine residues. It is worth noting, however, that there are probably many additional protein motifs used to recognize histone post-translational modification states. These include SANT domains, which recognize unmodified histone tails, and PhD domains, which recognize methylated histones.

5.5.1 Chromodomains

The name chromodomain derives from the fact that these domains were first discovered in two *Drosophila* proteins, Polycomb and HP1, that modulate transcription by regulating chromatin structure. These two factors play central roles in epigenetic transcriptional regulation and will be discussed extensively in Chapter 6.

Chromodomains bind methyl-lysine-containing histone tails with high specificity: the Polycomb chromodomain binds the histone H3 tail di- or trimethylated at lysine 27, while HP1 binds the histone H3 tail di- or trimethylated at lysine 9. The atomic resolution X-ray crystal structure of the Polycomb chromodomain bound to a portion of the H3 tail containing trimethyl-lysine 27 reveals the basis for recognition (Plate 5.2A). The trimethyl-lysine residue sits in a deep hydrophobic pocket lined with three aromatic amino acid side chains. The positively charged trimethyl-lysine is thought to make favorable electrostatic contact with the pi orbital electrons in the aromatic rings. The seven amino acid residues N-terminal to the trimethyl-lysine assume an extended conformation and thread through a channel in the chromodomain. Hydrogen bonds and apolar interactions between these N-terminal residues and the residues lining the channel explain the ability of the chromodomain to distinguish between different sites of methylation.

5.5.2 Bromodomains

Crystal structures of bromodomains bound to acetyl-lysine-containing peptides show that the acetyl-lysine side chain resides in a hydrophobic pocket in the bromodomain (Plate 5.2B). Residues C-terminal to the acetyl-lysine make sequence-specific and non-specific contacts in a shallow depression on the bromodomain extending from the pocket. Thus, different bromodomains have the capacity to distinguish between different acetyl-lysine residues depending upon sequence context. In general, bromodomains are less specific than are chromodomains. Accordingly, the contacts between bromodomains and acetylated peptides are not as extensive as the contacts between chromodomains and methylated peptides.

The bromodomain was first noticed in the *Drosophila* Brahma protein (and hence the name). Brahma is the catalytic subunit of the *Drosophila*

SWI/SNF complex. Indeed, a hallmark of the SWI/SNF subfamily ATPases is the presence of a bromodomain in addition to the ATPase domain. The presence of bromodomains in chromatin remodeling complexes suggests that histone acetylation leads to gene activation by recruiting such enzymes, which then catalyze changes in chromatin structure that render the template accessible to the transcriptional machinery (Figure 5.8).

In addition to being found in chromatin remodeling complexes, bromodomains are also frequently found in HATs, for example Gcn5 (see Table 5.1). This suggests that HATs are recruited to chromatin by the very modifications that they themselves generate. This could help to set up a positive feedback loop that strengthens and stabilizes an activated state (Figure 5.8).

5.6 COORDINATION OF ACTIVATOR–COACTIVATOR INTERACTIONS

In the last two chapters, we have seen that activator proteins orchestrate transcription by interacting with a myriad of coactivator proteins, including the Mediator, TAFs (TATA-binding protein-associated factors), P-TEFb (positive transcription elongation factor b), HAT complexes, histone ubiquitylating enzymes, and ATP-dependent chromatin remodeling complexes. Indeed, many factors such as Myc, Gal4, VP16, heat shock factor, and nuclear receptors are capable of interacting with members of many or all of these coactivator families. In many cases, a combination of biochemical and genetic studies suggest that all these interactions contribute to the ability of a single activator to stimulate transcription.

How are all these interactions coordinated with one another? Given the large size of many of these coactivator complexes, it seems unlikely that a single activator can recruit multiple coactivators simultaneously. Instead, sequential coactivator recruitment seems more likely and has been observed in studies of a number of promoters (e.g., see the studies of the IFNβ promoter described in Chapter 7). In one plausible scenario, activators could recruit a HAT complex first. The resulting acetylation of histones would provide docking sites for the bromodomain in Snf2. This might then lead to the cooperative recruitment of the SWI/SNF complex by simultaneous interactions between this complex, the activator, and histone acetyl-lysine residues. Mobilization or removal of nucleosomes by the SWI/SNF complex to expose the core promoter might then allow the activator to recruit the general machinery via interactions with the Mediator and TAFs leading to preinitiation complex assembly, open complex formation, and promoter clearance. Finally, the activator could recruit elongation factors such as P-TEFb to release Pol II from promoter proximal pause sites. There are, of course, many other plausible scenarios, and a variety of studies suggest that the order of coactivator recruitment is promoter specific.

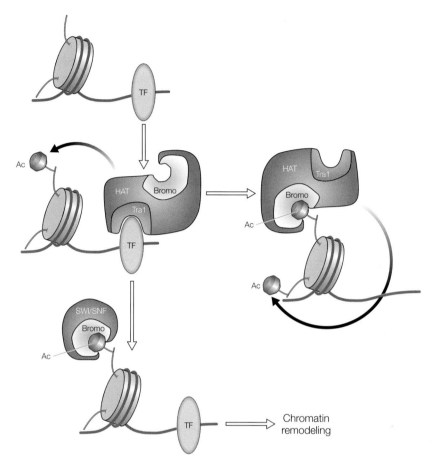

Figure 5.8 *Bromodomain function.* Bromodomains are often found in both ATP-dependent chromatin remodeling complexes such as the SWI/SNF complex and in histone acetyltransferase (HAT) complexes such as the SAGA complex. HATs are initially recruited to genes by sequence-specific transcription factors (TF). These interactions often involve the Tra1 protein (or its metazoan homolog TRRAP), which is a component of many HAT complexes. Acetylation can then result in recruitment of additional HAT complexes through interactions between the bromodomain and acetyl-lysine residues. This sets up a positive feedback loop. Acetylation can also result in the recruitment of ATP-dependent remodeling complexes, leading to chromatin remodeling and increased accessibility of the template to the transcriptional machinery.

5.7 SUMMARY

The DNA in eukaryotic cells is organized into a nucleoprotein complex called chromatin, a fact that has profound implications for gene regulation. The basic repeating unit of chromatin is the nucleosome, which consists of a histone octamer wrapped with DNA. Histone tails protrude from the nucleosome and together with linker histones mediate the higher order folding of chromatin into 30 nm chromatin fibers, chromonema fibers, and loops of chromonema fibers anchored to a protein scaffold.

Chromatin can be categorized as either euchromatin, which decondenses during interphase, or heterochromatin, which remains condensed throughout the cell cycle. The formation of constitutive heterochromatin at centromeres and telomeres is required for chromosome maintenance, while the formation of facultative heterochromatin in other regions of the genome leads to gene silencing.

The histone tails are intensively modified: the modifications include lysine acetylation, lysine and arginine methylation, serine phosphorylation, and lysine ubiquitylation. The combination of histone modifications present at any given gene is thought to encode the information that determines the transcriptional state of that gene. The enzymes that catalyze these modifications are often recruited to specific genes by protein–protein interactions with DNA-bound activators and repressors.

HATs and HDACs were discovered through yeast genetic screens to look for proteins with coactivator and corepressor activity. The finding that these coregulatory factors could catalyze histone acetylation and deacetylation provided the first direct link between transcriptional control and histone modification. While increased overall levels of histone acetylation generally lead to gene activation, the consequences of acetylating any particular histone residue depends on both the chromosomal and cellular environment.

Histone lysine methylation represents a relatively stable mark that may have a critical role in the epigenetic control of gene activity. HKMTs are highly specific enzymes that often only methylate a single histone lysine residue. Histone lysine demethylases are also highly specific and overcome the stability of the methylated lysine residue by employing oxidative mechanisms to release the methyl groups. Histone lysine methylation can lead to either gene activation or repression depending on the precise location of the methyl marks within the histone tails.

Histone modifications do not occur independently of one other. Rather, a programmed sequence of modification events is frequently required for gene activation or repression. For example, arginine methylation, serine phosphorylation, and lysine ubiquitylation are often required for subsequent histone modifications such as lysine acetylation and lysine methylation.

In addition to coregulators that covalently modify histones, the regulation of transcriptional state relies upon ATP-dependent chromatin remodeling enzymes, which catalyze non-covalent changes in chromatin structure. Like HATs and HDACs, these enzymes function in the context of multisubunit chromatin remodeling complexes, such as the SWI/SNF complex. The remodeling enzymes couple the hydrolysis of ATP to nucleosome sliding, changes in nucleosome structure, or the formation of nucleosome-free DNA. This is important to allow the transcriptional machinery access to the template and to facilitate processive transcriptional elongation. Targeting of chromatin remodeling complexes to DNA occurs by multiple mechanisms including direct interactions with activators, interactions with appropriately modified histones, and interactions with RNA polymerase.

Finally, the interpretation of the histone modification state depends on protein motifs that bind modified histone residues with high specificity. These motifs include chromodomains, which bind to methylated lysine residues, and bromodomains, which bind to acetylated lysine residues. Bromodomains are frequently found in both HAT complexes and ATP-dependent chromatin remodeling complexes. Thus, interactions between bromodomains and acetylated histones function to coordinate the activities of chromatin modifying and remodeling activities and to amplify the effects of histone acetylation.

PROBLEMS

1 The histone octamer is not stable in the absence of DNA. How does this fact contribute to the ability of chromatin remodeling complexes to disassemble nucleosomes?

2 The chromodomain in Polycomb recognizes the trimethylated form of histone H3 Lys27, while the chromodomain in HP1 recognizes the trimethylated form of histone H3 Lys9. What is so remarkable about the ability of these two chromodomains to distinguish between these two methylated lysine residues? How does the crystal structure of the chromodomain in complex with a methylated peptide help to explain this specificity?

3 The SAS complex, a MYST family HAT complex, may be the major histone H4 Lys16 acetyltransferase in yeast. SAS plays an important role in blocking the formation of heterochromatin. Does this role for SAS fit with the known relationship between histone H4 Lys16 acetylation and gene activity in yeast? Why or why not?

4 As discussed in this chapter, Gcn5 turned up in a screen for proteins required for the function of the VP16 activation domain. The screen was based on the observation that expression of high levels of a fusion protein consisting of the Gal4 DNA-binding domain fused to the VP16 activation domain is toxic to yeast cells. Cells carrying the Gal4-VP16 overexpression construct were mutagenized to look for mutations that would alleviate the toxicity. All the reported mutations mapped to the ADA genes, which encode Gcn5 plus additional components of the SAGA complex. What other genes might one expect to detect in such a screen? Why do you think mutants in these other genes were not recovered in this screen?

FURTHER READING

Chromatin structure

Luger, K., Mader, A.W., Richmond, R.K., Sargent, D.F. and Richmond, T.J. (1997) Crystal structure of the nucleosome core particle at 2.8 A resolution. *Nature*, **389**, 251–260. *The first near atomic resolution structure of the nucleosome core.*

Tumbar, T., Sudlow, G. and Belmont, A.S. (1999) Large-scale chromatin unfolding and remodeling induced by VP16 acidic activation domain. *J Cell Biol*, **145**, 1341–1354. *A study employing fluorescence microscopy showing that euchromatin takes the form of ~100 nm fibers in vivo and that heterochromatin is a further condensed form of these fibers. This paper also shows that activation domains can trigger the decondensation of heterochromatin.*

Horn, P.J. and Peterson, C.L. (2002) Molecular biology. Chromatin higher order folding–wrapping up transcription. *Science*, **297**, 1824–1827. *A succinct review on higher order chomatin structure and its role in regulating transcription.*

Zacharias, H. (1995) Emil Heitz (1892–1965): chloroplasts, heterochromatin, and polytene chromosomes. *Genetics*, **141**, 7–14. *A historical account of Emil Heitz's role in the discovery of heterochromatin.*

Histone modification

Histone modification code

Strahl, B.D. and Allis, C.D. (2000) The language of covalent histone modifications. *Nature*, **403**, 41–45. *The first use of the term "histone code" to describe the idea that the modification state of histones might constitute a code that determines the transcriptional state of each gene in a genome.*

Histone acetylation

Allfrey, V.G., Faulkner, R. and Mirsky, A.E. (1964) Acetylation and methylation of histones and their possible role in the regulation of RNA synthesis. *Proc Natl Acad Sci USA*, **51**, 786–794. Hebbes, T.R., Thorne, A.W. and Crane-Robinson, C. (1988) A direct link between core histone acetylation and transcriptionally active chromatin. *EMBO J*, **7**, 1395–1402. *Two papers demonstrating a correlation between histone acetylation state and transcriptional activity.*

Marcus, G.A., Silverman, N., Berger, S.L., Horiuchi, J. and Guarente, L. (1994) Functional similarity and physical association between GCN5 and ADA2: putative transcriptional adaptors. *EMBO J*, **13**, 4807–4815. Brownell, J.E., Zhou, J., Ranalli, T., Kobayashi, R., Edmondson, D.G., Roth, S.Y. and Allis, C.D. (1996) Tetrahymena histone acetyltransferase A: a homolog to yeast Gcn5p linking histone acetylation to gene activation. *Cell*, **84**, 843–851. *Two papers that together strongly suggest a cause-and-effect relationship between histone acetylation and gene activation. The first presents the characterization of a Gcn5-containing complex that is required for activation by a number of sequence-specific activation, while the second demonstrates that Gcn5 is a histone acetyltransferase.*

Grant, P.A., Duggan, L., Cote, J., Roberts, S.M., Brownell, J.E., Candau, R., Ohba, R., Owen-Hughes, T., Allis, C.D., Winston, F., Berger, S.L. and Workman, J.L.

(1997) Yeast Gcn5 functions in two multisubunit complexes to acetylate nucleosomal histones: characterization of an Ada complex and the SAGA (Spt/Ada) complex. *Genes Dev*, **11**, 1640–1650. *The first detailed characterization of the SAGA complex.*

Taunton, J., Hassig, C.A. and Schreiber, S.L. (1996) A mammalian histone deacetylase related to the yeast transcriptional regulator Rpd3p. *Science*, **272**, 408–411. *The first demonstration of a direct link between histone deacetylation and repression.*

Laherty, C.D., Yang, W.M., Sun, J.M., Davie, J.R., Seto, E. and Eisenman, R.N. (1997) Histone deacetylases associated with the mSin3 corepressor mediate mad transcriptional repression. *Cell*, **89**, 349–356. McMahon, S.B., Wood, M.A. and Cole, M.D. (2000) The essential cofactor TRRAP recruits the histone acetyltransferase hGCN5 to c-Myc. *Mol Cell Biol*, **20**, 556–562. *Two papers showing that Myc family bHLH factors direct activation and repression by recruiting histone acetyltransferases and histone deacetylases.*

Kurdistani, S.K., Tavazoie, S. and Grunstein, M. (2004) Mapping global histone acetylation patterns to gene expression. *Cell*, **117**, 721–733. Akhtar, A. and Becker, P.B. (2000) Activation of transcription through histone H4 acetylation by MOF, an acetyltransferase essential for dosage compensation in Drosophila. *Mol Cell*, **5**, 367–375. *Two papers that together demonstrate the context-dependent nature of the effects of histone acetylation on transcription. The first paper shows that histone H4 lysine 16 acetylation most often correlates with repression, while the second paper shows that acetylation of this lysine leads to the upregulation of gene expression in dosage compensation.*

Shogren-Knaak, M., Ishii, H., Sun, J.M., Pazin, M.J., Davie, J.R. and Peterson, C.L. (2006) Histone H4-K16 acetylation controls chromatin structure and protein interactions. *Science*, **311**, 844–847. *In vitro analysis of the effect of histone acetylation on higher order chromatin structure.*

Histone methylation

Rea, S., Eisenhaber, F., O'Carroll, D., Strahl, B.D., Sun, Z.W., Schmid, M., Opravil, S., Mechtler, K., Ponting, C.P., Allis, C.D. and Jenuwein, T. (2000) Regulation of chromatin structure by site-specific histone H3 methyltransferases. *Nature*, **406**, 593–599. *The first demonstration that the SET domain catalyzes the methylation of histone lysine residues.*

Shi, Y., Lan, F., Matson, C., Mulligan, P., Whetstine, J.R., Cole, P.A., Casero, R.A. and Shi, Y. (2004) Histone demethylation mediated by the nuclear amine oxidase homolog LSD1. *Cell*, **119**, 941–953. Tsukada, Y., Fang, J., Erdjument-Bromage, H., Warren, M.E., Borchers, C.H., Tempst, P. and Zhang, Y. (2006) Histone demethylation by a family of JmjC domain-containing proteins. *Nature*, **439**, 811–816. *Two of the first papers documenting the existence of histone demethylases. The first presents the characterization of a class I demethylase, while the second presents the characterization of a class II demethylase.*

Cross talk between histone modifications

Huang, S., Litt, M. and Felsenfeld, G. (2005) Methylation of histone H4 by arginine methyltransferase PRMT1 is essential in vivo for many subsequent histone modifications. *Genes Dev*, **19**, 1885–1893. *Cross talk between arginine methylation and lysine acetylation.*

Lo, W.S., Duggan, L., Emre, N.C., Belotserkovskya, R., Lane, W.S., Shiekhattar, R. and Berger, S.L. (2001) Snf1 – a histone kinase that works in concert with the histone acetyltransferase Gcn5 to regulate transcription. *Science*, **293**, 1142–1146. *Cross talk between serine phosphorylation and lysine acetylation.*

Kao, C.F., Hillyer, C., Tsukuda, T., Henry, K., Berger, S. and Osley, M.A. (2004) Rad6 plays a role in transcriptional activation through ubiquitylation of histone H2B. *Genes Dev*, **18**, 184–195. *Cross talk between ubiquitylation, deubiquitylation, and lysine methylation.*

ATP-dependent chromatin remodeling

Peterson, C.L. and Herskowitz, I. (1992) Characterization of the yeast SWI1, SWI2, and SWI3 genes, which encode a global activator of transcription. *Cell*, **68**, 573–583. Imbalzano, A.N., Kwon, H., Green, M.R. and Kingston, R.E. (1994) Facilitated binding of TATA-binding protein to nucleosomal DNA. *Nature*, **370**, 481–485. Kwon, H., Imbalzano, A.N., Khavari, P.A., Kingston, R.E. and Green, M.R. (1994) Nucleosome disruption and enhancement of activator binding by a human SW1/SNF complex. *Nature*, **370**, 477–481. *Three papers demonstrating the existence of the chromatin remodeling SWI/SNF complex. The first paper presents a genetic analysis showing that the SWI genes encode components of a complex generally required for activation, while the second and third papers show that the ATP-dependent activity of this complex is required for the binding of sequence specific transcription factors to a chromatin template.*

Hamiche, A., Sandaltzopoulos, R., Gdula, D.A. and Wu, C. (1999) ATP-dependent histone octamer sliding mediated by the chromatin remodeling complex NURF. *Cell*, **97**, 833–842. *The characterization of the chromatin sliding activity of an ISWI family chromatin remodeling complex.*

Recognition of the histone modification state

Hassan, A.H., Prochasson, P., Neely, K.E., Galasinski, S.C., Chandy, M., Carrozza, M.J. and Workman, J.L. (2002) Function and selectivity of bromodomains in anchoring chromatin-modifying complexes to promoter nucleosomes. *Cell*, **111**, 369–379. *Demonstration that bromodomains in chromatin remodeling complexes can direct these complexes to acetylated chromatin.*

Fischle, W., Wang, Y., Jacobs, S.A., Kim, Y., Allis, C.D. and Khorasanizadeh, S. (2003) Molecular basis for the discrimination of repressive methyl-lysine marks in histone H3 by Polycomb and HP1 chromodomains. *Genes Dev*, **17**, 1870–1881. *High resolution structure of the complex between a chromodomain and a methyl-lysine-containing histone peptide.*

Owen, D.J., Ornaghi, P., Yang, J.C., Lowe, N., Evans, P.R., Ballario, P., Neuhaus, D., Filetici, P. and Travers, A.A. (2000) The structural basis for the recognition of acetylated histone H4 by the bromodomain of histone acetyltransferase gcn5p. *EMBO J*, **19**, 6141–6149. *High resolution structure of the complex between a bromodomain and an acetyl-lysine-containing histone peptide.*

6

Epigenetic control of transcription

Key concepts

- Transcriptional states can be remarkably stable and therefore heritable
- Heritable transcriptional states, often termed epigenetic states, may be as important as genetic states in determining phenotype
- Histone methylation and the recognition of methylated histones by chromodomain factors are central to the maintenance of epigenetic states
- Non-coding RNAs often play key roles in the initiation of epigenetic states

6.1 INTRODUCTION

As discussed in the last chapter, covalent and non-covalent changes in chromatin structure profoundly influence the transcriptional state in eukaryotes. As will be discussed in this chapter, the different forms of chromatin associated with different transcriptional states are sometimes remarkably stable, and can be preserved during replication of the chromosomes and cell division. Transcriptional states can therefore be transmitted from cell generation to cell generation. This phenomenon is termed epigenetic ("near genetic") inheritance, because the transcriptional states and the phenotypes associated with them behave, in many respects, as if they were conveyed by differences in the DNA sequence.

Phenotype may depend as much on epigenetics as it does on genetics. Genome-wide analysis of chromatin structure in genetically identical individuals (monozygotic twins) suggests that differences in disease susceptibility or personality may be due to epigenetic differences (Box 6.1). Furthermore, epigenetics is a major factor in the development of multicellular organisms, in which undifferentiated embryos give rise to organisms composed of diverse tissues. With minor exceptions, all tissues in a single organism are genetically identical, and yet each tissue possesses distinctive characteristics. The unique identity of each tissue results from a unique combination of epigenetic states.

This chapter will explore several examples of epigenetic inheritance. The first is heterochromatic silencing, which is found to exhibit remarkable epigenetic stability. As discussed in the last chapter, heterochromatin is a highly condensed form of chromatin, and is of two types, constitutive heterochromatin and facultative heterochromatin. While constitutive heterochromatin is associated with centromeres and telomeres, which are largely devoid of genes, facultative heterochromatin functions to stably repress genes over multiple cell generations.

The discussion of heterochromatic silencing will be followed by a presentation of the mechanisms by which two groups of gene products, the Trithorax group (TrxG) and the Polycomb group (PcG), maintain epigenetically stable transcriptional states. Both TrxG proteins, which are responsible for the maintenance of active states, and PcG proteins, which are responsible for the maintenance of silent states, ensure that such states are faithfully transmitted from cell generation to cell generation throughout the development of an organism. As will become clear, similar strategies are used to achieve heritable transcriptional states in both heterochromatic silencing and PcG silencing.

Finally, the chapter will conclude with a brief discussion of X chromosome inactivation, in which the epigenetic silencing of one X chromosome in each cell of the mammalian female is essential to compensate for the difference in X-linked gene dosage between males and females. The inactive X chromosome exhibits hallmarks of both facultative

Box 6.1

What does epigenetics have to say about who we are?

Genetically identical individuals such as identical twins can have remarkably different phenotypes. Many of us may be acquainted with twins who have distinctive personalities, likes, and dislikes. Less subjectively, studies of disease occurrence often show surprisingly low identical twin concordance rates. For example, when one identical twin is born with a cleft palate, the chance of the other twin being similarly afflicted is only about 10%. Similarly, the twin concordance rates for type 1 diabetes and breast cancer are only about 35% and 10%, respectively. Classically, this variation has been attributed to the influence of environment, and it is certainly easy to understand how environmental factors such as nutrition, teratogens, or carcinogens could influence rates of diabetes, birth defects, or cancer. In the last 10 years or so, however, it has become increasingly clear that epigenetic variability may contribute as much to this phenotypic variation as does environmental variability (Figure B6.1).

Genome-wide analysis shows that chromosomal modifications such as cytosine methylation or histone acetylation associated with changes in epigenetic state can differ dramatically between two identical twins, and these differences increase with age. While such studies suggest that epigenetic differences accumulate in somatic tissues over time, many findings also provide direct evidence that epigenetic differences in multicellular organisms can be inherited through the germ line. Two examples of germ line inheritance of epigenetic states to be presented below, one in plants and the other in humans, display certain parallels. In both examples, genes that are normally active have adopted a silent state. Although the exact cause of the silencing is not known, the silenced genes exhibit increased levels of cytosine methylation in their 5′ flanking regions, a modification often linked to epigenetic silencing. These examples suggest

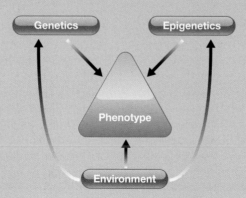

Figure B6.1 *Phenotype is a product of genetic, epigenetic, and environmental influences.*

that our identities may depend as much upon epigenetics as upon genetics and the environment.

Inactivation of the SUPERMAN gene in *Arabidopsis thaliana* results in flowers with extra stamens. One set of SUPERMAN variants (the "*clark kent* alleles") display heritable decreases in the level of SUPERMAN expression. These variants are not mutant alleles in the normal sense of the term since they do not exhibit changes in the DNA sequence. The decreased level of expression is instead due to epigenetic silencing, and to denote this fact the *clark kent* alleles are termed "epi-alleles". Consistent with the idea of epigenetic silencing, increased cytosine methylation is detected in the 5′ flanking regions of the *clark kent* alleles.

Another example of epigenetic inheritance is provided by studies of a hereditary form of colon cancer known as hereditary non-polyposis colon cancer (HNPCC). This disease is often linked to defects in mismatch repair, a process that corrects occasional mistakes introduced into the DNA during replication. In some HNPCC patients, the disease results from epigenetic silencing of the mismatch repair gene *MSH2*. In these individuals, the silent epi-allele is marked by cytosine methylation of the promoter, which can be passed from parent to offspring through the gametes (Figure B6.2).

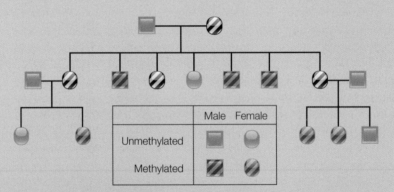

Figure B6.2 *A pedigree of a family afflicted with the hereditary form of colon cancer known as HNPCC.* In this family, the disease is thought to result from the epigenetic silencing of the DNA repair gene *MSH2*. Individuals in which this gene is transcriptionally silent exhibit extensive cytosine methylation in the promoter of the gene. The family tree shows that this methylated state can be passed on from generation to generation through the germ line. Individuals afflicted with HNPCC (shown as a black and white stripe) have all inherited the methylated epi-allele. It is possible that some of the individuals who are not thus far afflicted, will develop HNPCC at a later date since this disease typically manifests itself in middle age.

heterochromatin and PcG-silenced genes further highlighting the connections between heterochromatic and PcG silencing.

6.1.1 Common themes in epigenetics: a central role for histone methylation

This chapter highlights a number of recurring themes in the epigenetic regulation of transcription. These are as follows:

1 **Histone methylation is used to mark both active and silent epigenetic states.** Many years before the role of histone covalent modification in transcriptional regulation was appreciated, *Drosophila* molecular geneticists noticed a conserved motif termed the SET domain in many proteins required for the inheritance of transcriptional states. Indeed, SET is an acronym for three gene products (Su(var) 3-9, Enhancer of zeste, and Trithorax) that will be featured prominently in this chapter. The subsequent discovery that the SET domain is a catalytic domain mediating histone lysine methylation (see Chapter 5) provided a solid connection between histone methylation and epigenetics.

2 **Chromodomain-containing factors are used to recognize methylation marks and interpret them appropriately.** Chromodomains bind to methylated lysine residues in a highly specific and context-dependent manner (see Chapter 5). As a result, chromodomains are readily able to direct different regulatory outcomes in response to different methylated lysine residues in the histone tails.

3 **The initial formation of epigenetic states occurs by two general mechanisms.** First, initiation frequently requires non-coding RNAs, which recruit the heterochromatin-forming machinery (e.g., histone lysine methyltransferases) to appropriate loci. Second, initiation also utilizes sequence-specific DNA-binding proteins to recruit the heterochromatin-forming machinery.

4 **Positive feedback loops are used to reinforce and stabilize epigenetic states.** For example, epigenetic histone modifications may mediate their own renewal by helping to recruit the enzymes that catalyze their formation.

6.2 HETEROCHROMATIC SILENCING

Much of what we know about the mechanisms behind heterochromatic silencing has come from genetic and biochemical analysis of the fission yeast *Schizosaccharomyces pombe*. Therefore, this section will focus largely on studies carried out using this particular model organism. As will also be discussed in this section, the findings from studies of *S. pombe* are applicable to most eukaryotes.

6.2.1 Chromosomal inheritance of the silent state

Studies of heterochromatic silencing in *S. pombe* have focused, in part, on the role of heterochromatin in the regulation of the mating-type loci. *S. pombe* has two different mating types termed plus (P) and minus (M). Haploid cells are capable of growth by cell division (vegetative growth), and can also fuse with cells of the opposite mating type to produce diploid cells. Diploids can grow vegetatively if supplied with adequate nutrients or will undergo meiosis and sporulation when deprived of adequate nutrients. The haploid spores can then germinate and commence vegetative growth when nutrients again become available. However, the haploid is less robust than the diploid and is also unable to sporulate if nutrients become scarce. Consequently, it is advantageous for cells to switch mating type at high frequency thereby making it possible for the descendents of a single spore to fuse and produce diploids.

Meiosis – the process in which the chromosomal content of a cell is reduced from two copies to one copy of each chromosome. Meiosis involves two cell divisions producing four haploid cells from a single diploid cell. Prior to the first division, the chromosomes replicate so that each chromosome consists of two identical, tightly associated sister chromatids. During the first division, the homologous chromosomes separate, while, during the second cell division, the sister chromatids separate.

Mating type in *S. pombe* is dictated by the *mat1* locus, which can contain either the P cassette, producing a P mating-type cell, or the M cassette, producing an M mating-type cell. Two additional loci, *mat2* and *mat3*, contain additional cassettes of mating-type information; *mat2* contains the P cassette, while *mat3* contains the M cassette (Figure 6.1). *mat2* and *mat3* are normally silent – they exist solely to allow for the possibility

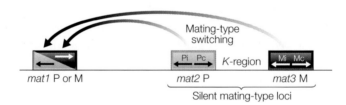

Figure 6.1 *The* S. pombe *mating-type loci.* The *S. pombe* genome contains three mating-type loci, all located on the same chromosome. *mat1* is transcriptionally active. If it contains the P cassette, the cell is of the P mating type, while if it contains the M cassette, the cell is of the M mating type. *mat2*, which contains the P cassette, and *mat3*, which contains the M cassette, are transcriptionally silent due to the silencer located in the intervening *K*-region. Mating-type switching occurs when copies of *mat2* or *mat3* replace the sequences at *mat1*.

of mating-type switching. A P mating-type cell switches to the M mating type when the M cassette in *mat3* gives rise to a copy of itself that replaces the P cassette at *mat1*. Similarly, an M mating-type cell switches to the P mating type when the P cassette in *mat2* gives rise to a copy of itself that replaces the M cassette at *mat1*.

Expression of the mating-type information at *mat2* and *mat3* would block mating since the cell would be expressing both P and M mating-type information. Therefore, *mat2* and *mat3* must be maintained in a silent state. The transcriptional silencing of *mat2* and *mat3* depends upon the 11 kb region between these two loci termed the *K*-region, which functions as a silencer. Silencing requires a number of trans-acting factors, including those encoded by the *clr* (cryptic loci regulator) genes as well as by *swi6*. As will be discussed below, several of these genes encode histone modifying enzymes, thus implying a link between histone modifications and silencing. These trans-acting factors are also required for the silencing of genes inserted into constitutive heterochromatin (e.g., centromeric heterochromatin), strongly suggesting that the silencing of the mating-type loci involves the organization of these loci into a facultative heterochromatic domain.

The silent heterochromatic state is epigenetically stable. This is best shown by studies in which genetic markers such as the *ura4* gene are inserted into the *K*-region or into centromeric chromatin. The Ura4 protein catalyzes a step in the synthesis of uracil and thus cells lacking an active copy of *ura4* are unable to grow in medium lacking uracil (i.e., they exhibit a Ura⁻ phenotype). Normally, a gene inserted into the *K*-region (or into centromeric chromatin) would be incorporated into heterochromatin and would therefore be silent. However, partial deletions of the *K*-region can sometimes cripple the silencer. Rather than partially alleviating silencing in all the cells, these deletions often lead to variegation, meaning the *ura4* gene inserted into the crippled *K*-region will be completely active in some of the cells and completely silent in the remaining cells (Box 6.2).

> **Variegation** – this term was originally used to describe plants with different colored zones in the leaves. It can be readily extended to fly eyes with differently colored patches of tissue or to any organism or colony of microbes in which genetically identical cells exhibit a mixture of different phenotypes.

An important feature of the active and silent states in strains exhibiting such variegation is their heritability – that is, most cells remain in the same state through multiple rounds of cell division. In the case of variegated cells containing the *ura* marker gene in the *K*-region, the vast majority of the descendents of a Ura⁺ cell maintain the Ura⁺ phenotype, while the vast majority of the descendents of a Ura⁻ cell maintain the Ura⁻ phenotype. Thus, the silent (heterochromatic) and active (euchromatic) states are metastable states that can be transmitted from cell generation to cell generation. The fact that cells do

Box 6.2

Mendelian inheritance of epigenetic states

An excellent way to determine if a trait is chromosomally inherited is to determine how the trait segregates during meiosis, the process that converts a diploid cell into four haploid cells. Meiosis begins with DNA replication (Figure B6.2). The replicated chromosomes then pair with their homologs in a process termed synapsis. This is followed by the first meiotic division during which homologous chromosomes separate, and then the second meiotic division during which sister chromatids separate. In yeast such as *S. pombe*, the four haploid cells resulting

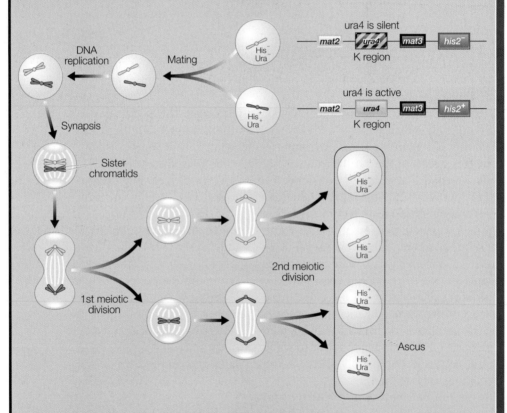

Figure B6.2 *Meiotic segration of epigenetic states.* This diagram illustrates the results of mating of a fission yeast strain containing a *K*-region *ura4* insert in the silent epigenetic state to a fission yeast strain containing a *K*-region *ura4* insert in the active epigenetic state. The Ura⁺ and Ura⁻ phenotypes always segrate in a 2 : 2 ratio indicating that the epigenetic states are chromosomally inherited. Cosegregation of the Ura⁺ phenotype with the His⁺ phenotype shows that the epigenetic states are closely linked to the *K*-region.

from these two cell divisions are contained within a spore case called an ascus. This allows unambiguous identification of the four haploid progeny arising from a single premeiotic diploid. In an experimental technique termed tetrad dissection, the ascus is removed and the four haploid cells are carefully manipulated (using a "micromanipulator") into labeled positions on an agar plate. The haploid cells are allowed to grow into colonies, which are then examined to determine their phenotype. Most often, this involves transferring samples of each colony onto different types of growth medium to determine their nutritional requirements.

To prove that the epigenetic state of the *K*-region is chromosomally inherited, two variegated haploid strains carrying insertions of the *ura4* gene in the *K*-region were crossed with one another (Figure B6.2). In one of the strains, *ura4* was in the silent epigenetic state (the strain exhibited the Ura⁻ phenotype), while in the other strain, *ura4* was in the active epigenetic state (the strain exhibited the Ura⁺ phenotype). The Ura⁻ parent also contained a mutation in the *his2* gene, which encodes an enzyme required for histidine biosynthesis, and which is very close to the *K*-region. The diploid cells resulting from the cross were allowed to undergo meiosis and multiple tetrads were isolated and dissected. Plating of the haploid colonies onto media lacking uracil showed that two of the meiotic products in each tetrad were Ura⁺ and two were Ura⁻. The 2 : 2 segregation ratio is exactly what Mendelian genetics predicts for a trait conferred by a single gene and thus this result proves chromosomal inheritance of the epigenetic state. (What ratio might you expect for a cytoplasmically inherited trait?) Plating of the same haploid colonies onto media lacking histidine showed that the Ura⁻ segregants were invariably His⁻, proving that the Ura⁻ epigenetic state segregates with the *K*-region inherited from the original Ura⁻ parent.

occasionally switch states shows that the states are epigenetic rather than genetic states.

Throughout this chapter, epigenetic states will be equated with heritable chromatin structures. However, an alternative to chromosomal inheritance, namely cytoplasmic inheritance, can also account for epigenetic phenomena. In cytoplasmic inheritance, cells contain cytoplasmic determinants that are responsible for a particular phenotype. Cells containing the determinant exhibit one phenotype, while cells lacking the determinant exhibit a different phenotype. If such a determinant is divided between the two daughter cells during cell division, the daughter cells will both inherit the phenotype associated with the determinant.

How do we know then that the active and silent states in variegated strains containing modified *K*-regions are due to heritable chromatin structures rather than to cytoplasmic determinants? The answer to this question is that the two states obey the laws of Mendelian inheritance, a finding that is compatible with chromosomal inheritance, but not with cytoplasmic inheritance (Box 6.2).

6.2.2 Histone methylation and maintenance of the silent state

What is the molecular basis for the inheritance of the transcriptionally silent heterochromatic state? This single question can be subdivided into two questions, which will be addressed in reverse order:

1 What are the mechanisms that initiate formation of this epigenetic state?
2 What are the mechanisms responsible for the maintenance of this state and its transmission from cell generation to cell generation?

A partial answer to the second question is provided by analysis of the trans-acting factors required for heterochromatic silencing in *S. pombe*, including Clr3, Clr4, Clr6, and Swi6. One of the most critical of these factors is Clr4, which contains a SET domain. SET domains catalyze lysine methylation (see Chapter 5), and the Clr4 SET domain is exquisitely specific for histone H3 lysine 9, primarily mediating the formation of di- and trimethylated forms of this residue. Not surprisingly, therefore, heterochromatic regions (including centromeric heterochromatin and the silent mating-type loci) are highly enriched for di- and trimethylated histone H3 lysine 9, which serves as a silencing mark that can be transmitted during mitosis and meiosis.

Methyl-lysine recognition by the chromodomain factor Swi6

How does Swi6 direct hetero-chromatin formation?

As mentioned in the text, one unproven possibility is that Swi6 recruits histone deacetylases and that the resulting deacetylation of histone lysine residues leads directly to chromatin condensation. An alternative possibility is that Swi6 directly mediates interactions between nucleosomes, stabilizing a condensed form of chromatin.

How does histone H3 lysine 9 methylation direct heterochromatin maintenance? A key factor in this process is Swi6. Like many proteins involved in epigenetic transcriptional control, Swi6 contains a chromodomain, which mediates context-specific binding to methyl-lysine side chains (see Chapter 5). The Swi6 chromodomain is specific for the di- and trimethylated forms of histone H3 lysine 9, and thus methylation of histone H3 lysine 9 by Clr4 leads to the recruitment of Swi6. Swi6 can self-associate and this self-association may lead to the spreading of Swi6 along the chromosome (Figure 6.2). By mechanisms that we do not understand, Swi6 then stabilizes transcriptionally silent heterochromatin.

Histone deacetylation in heterochromatic silencing

While histone H3 lysine 9 methylation and Swi6 recruitment are essential for maintenance of the heterochromatic state, histone deacetylation

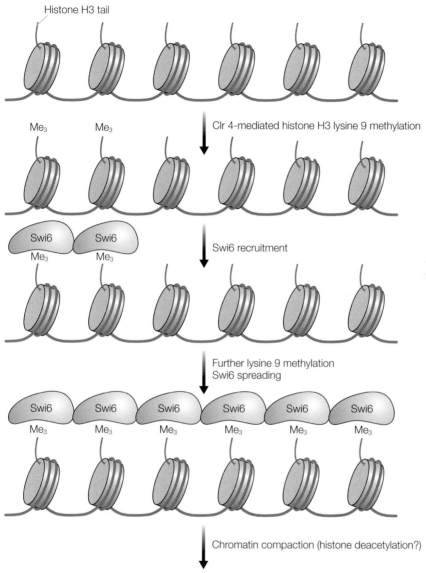

Histone H3 tail

Me₃   Me₃   Clr 4-mediated histone H3 lysine 9 methylation

Swi6   Swi6   Swi6 recruitment
Me₃   Me₃

Swi6   Swi6   Swi6   Swi6   Swi6   Swi6
Me₃   Me₃   Me₃   Me₃   Me₃   Me₃

Further lysine 9 methylation
Swi6 spreading

Chromatin compaction (histone deacetylation?)

Transcriptionally silent heterochromatin

Figure 6.2 *Model for heterochromatic silencing.* Heterochromatic silencing requires histone H3 lysine 9 di- and trimethylation, which is catalyzed by Clr4. The methylated lysine 9 residue (shown here in its trimethylated form) then serves as a docking site for the chromodomain factor Swi6. Lysine 9 methylation spreads by unknown mechanisms. Swi6 associates with itself and with the methylated histones to spread along the chromatin and establish a heterochromatic domain. It is not understood what drives chromatin compaction, but one possibility is that Swi6 recruits histone deacetylases such as Clr3, which deacetylate chromatin, favoring higher order folding of the chromatin fiber.

is also essential. Several histone deacetylases including the products of the *clr3* and *clr6* genes as well as an NAD$^+$-dependent histone deacetylase termed Sir2 (see the section below on heterochromatic silencing in budding yeast) are required for optimal heterochromatic silencing. Not surprisingly, therefore, heterochromatin exhibits reduced acetylation relative to euchromatin.

We do not understand how histone deacetylation fits into the overall pathway for heterochromatic silencing. However, as mentioned previously (Chapter 5), histone acetylation disrupts higher order chromatin structure. Thus, one intriguing although unproven possibility is that Swi6 favors the heterochromatic state by recruiting histone deacetylases. By removing acetyl groups from chromatin, these enzymes could drive chromatin compaction thereby promoting heterochromatic silencing.

6.2.3 Initiation of heterochromatic silencing

While the Clr proteins and Swi6 are required for the maintenance of heterochromatin, they are insufficient to initiate the formation of a heterochromatic domain. There appear to be two independent mechanisms for initiating heterochromatin formation: one involves the RNA interference machinery and the other involves sequence-specific DNA-binding proteins.

Initiation by double-stranded RNA

In RNA interference (RNAi), double-stranded RNA interferes with multiple steps in the gene expression pathway. The double-stranded RNAs that initiate this process often arise naturally during the replication of viruses or transposons and therefore the RNAi pathway serves as an intracellular defense against these parasites. All arms of the RNAi pathway (Figure 6.3A) require the cleavage of double-stranded RNA into short interfering RNAs (siRNAs) of 21–28 bp in length by the RNase III family endonuclease Dicer. In the best-understood branch of the pathway, one strand of the siRNA binds the RNA-induced silencing complex (RISC), which then uses the siRNA strand as a guide to find and degrade complementary transcripts. Argonaut, the core component of RISC, harbors both the siRNA binding and RNase activity of the complex.

In addition to triggering the degradation of mRNA, siRNA also triggers the formation of transcriptionally silent heterochromatin (Figure 6.3A). Centromeric DNA consists of repeat sequences ranging from a few basepairs to a few thousand basepairs in length (the length is species dependent), and transcription of these repeats gives rise to double-stranded RNA. This could be due to bidirectional transcription of the repeats or to the conversion of single-stranded RNA to double-stranded RNA by RdRP. Dicer then cleaves the double-stranded RNA into siRNA, which associates

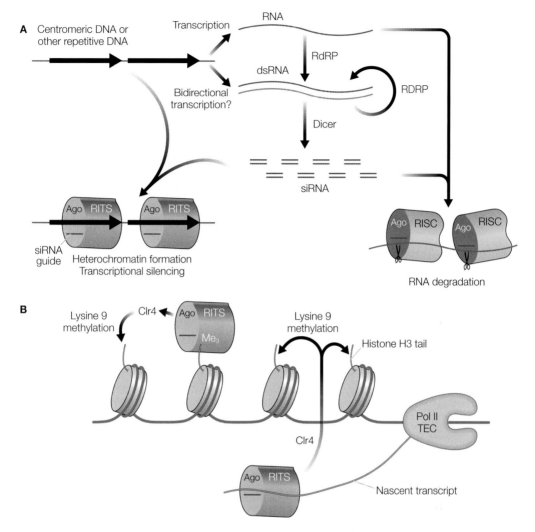

Figure 6.3 *The RNAi pathway and heterchromatin formation.* (A) The RNAi pathway. Transcription of repetitive DNA such as centromeric DNA results in the formation of double-stranded RNA. This may involve bidirectional transcription of the repeats as well as conversion of single-stranded RNA to double-stranded RNA by RNA-dependent RNA polymerase (RdRP). RdRP also catalyzes the amplification of double-stranded RNA. The endonuclease Dicer cleaves the double-stranded RNA into siRNA, which consists of 21–28 bp long fragments with two nucleotide 3′ overhangs. The siRNA is then recognized by Ago, which is a component of both the RITS (RNA-induced initiation of transcriptional gene silencing) complex and RISC (RNA-induced silencing complex). RISC directs RNA degradation (a form of post-transcriptional gene silencing), while the RITS complex directs heterochromatin formation. (B) Two mechanisms for targeting of RITS to chromatin. RITS may bind to nascent transcripts through a basepairing interaction with the siRNA. The RITS can then recruit Clr4 leading to histone H3 lysine 9 di- and trimethylation. In addition, RITS contains a subunit termed Chp1 that possesses a chromodomain. This chromodomain binds specifically to methylated lysine 9 providing another mechanism of recruiting RITS to heterochromatin. RITS recruited in this way can also bring in Clr4 leading to further lysine 9 methylation and therefore setting up a positive feedback loop. TEC, ternary elongation complex.

with a complex known as RITS (RNA-induced initiation of transcriptional gene silencing), a second Argonaut-containing complex that is distinct from RISC. Once loaded with centromeric siRNA, RITS associates with centromeric chromatin perhaps through basepairing between the siRNA in RITS and nascent centromeric transcripts (Figure 6.3B). RITS then initiates heterochromatin formation possibly by recruitment of Clr4, leading to histone H3 lysine 9 di- and trimethylation, Swi6 recruitment, and chromatin compaction.

In addition to siRNA-guided targeting of RITS, which is required for the initiation of heterochromatin formation, there appears to be an siRNA-independent mechanism for targeting RITS that may be required for the maintenance of heterochromatin (Figure 6.3B). This second mechanism relies on a subunit of RITS termed Chp1, which bears a chromodomain that binds di- and trimethylated histone H3 lysine 9. Thus, RITS can be directly recruited by heterochromatin and RITS can in turn direct heterochromatin formation. This positive feedback loop may help to set up a self-reinforcing state that is able to survive replication of the chromosomes.

Initiation by transcription factors

While the RNAi pathway is essential for initiation of the silent state at the centromeres, it is not absolutely required for silencing at the *S. pombe* silent mating-type loci. This indicates that there must be RNA-independent mechanisms for targeting the heterochromatin-forming machinery to these loci. In addition to sequences with extensive homology to the centromeres (which direct RNAi-dependent silencing, see Box 6.2), the *K*-region contains sequences that bind two closely related basic leucine zipper (bZIP) domain-containing sequence-specific transcription factors termed Atf1 and Pcr1. These factors are both capable of recruiting Clr4 and Swi6; the subsequent spreading of Swi6 then leads to the formation of the heterochromatic domain. Thus, sequence-specific transcription factors function independently of the RNAi machinery to direct heterochromatin formation at the silent mating-type loci.

6.2.4 Evolutionary conservation of mechanisms for heterochromatic silencing

Although the foregoing discussion has focused on findings made in *S. pombe*, the mechanisms used to initiate and maintain heterochromatic silencing appear to be highly conserved in the eukaryotic domain. For example, Clr4 and Swi6 homologs are involved in heterochromatic silencing in organisms as diverse as flowering plants, mammals, and insects. The use of the RNAi pathway to initiate heterochromatin formation seems to be similarly widespread.

Some of the strongest evidence that mechanisms of heterochromatic silencing are highly conserved comes from studies of a phenomenon in *Drosophila melanogaster* known as position effect variegation (PEV). These studies involve the *mottled 4* allele of *white* (*white^{m4}*), a gene that is required for the formation of red eye pigment. While wild-type flies have red eyes, mutations that inactivate *white* result in pure white eyes. *white^{m4}* is not a mutant allele in the normal sense of the term – rather it is a chromosomal inversion that places *white* abnormally close to a centromere (Plate 6.1A). Variable spreading of heterochromatin from the centromere results in the variable silencing of *white* and therefore in variegated eyes. Most importantly, the pigmentation does not vary in a cell-by-cell manner – instead the eyes contain large patches of pigmented cells, in which *white* is active, and large patches of unpigmented cells, in which *white* is silent (Plate 6.1B). This suggests that once the active (euchromatic) or silent (heterochromatic) state is established early in development, this state is maintained throughout the remaining rounds of cell division leading to the adult fly. The large pigmented and unpigmented patches in the eye represent clones of cells descended from a single progenitor in which the transcriptional state of *white* was faithfully transmitted from cell generation to cell generation until proliferation of the eye field was complete. Thus, just as in fission yeast, the euchromatic and heterochromatic states in fruit flies are subject to epigenetic inheritance.

Drosophila geneticists have sought for decades to understand epigenetic inheritance by screening for second site mutations that modify the eye pigmentation phenotype in *white^{m4}* flies. The dozens of genes identified in these studies have been grouped into two classes – the suppressors of variegation (*Su(var)* genes) and the enhancers of variegation (the *E(var)* genes) (Plate 6.1). Mutations in *E(var)* genes result in increased heterochromatic silencing of *white^{m4}* (more unpigmented eye tissue), implying that they encode proteins that favor euchromatin. Conversely, mutations in *Su(var)* genes result in decreased heterochromatic silencing (less unpigmented eye tissue), implying that they encode proteins that favor heterochromatin.

Two of the suppressors of variegation, *Su(var)3-9* and *Su(var)2-5*, are of particular interest here because they encode homologs of two factors that play pivotal roles in heterochromatic silencing in *S. pombe*. *Su(var)3-9* encodes a SET domain-containing histone H3 lysine 9 methyltransferase, which is the *Drosophila* counterpart of Clr4; *Su(var)2-5* encodes a chromodomain-containing protein commonly referred to as heterochromatin protein-1 (HP1), which is the *Drosophila* counterpart of Swi6. Just like the Swi6 chromodomain, the HP1 chromodomain binds to the di- and trimethylated forms of histone H3 lysine 9. Thus, studies of PEV in *Drosophila* provide compelling evidence that histone lysine 9 methylation is an evolutionarily conserved epigenetic modification for heterochromatic silencing.

6.2.5 DNA methylation and heterochromatin

This discussion of heterochromatic silencing has focused on histone H3 lysine 9 di- and trimethylation as a near-universal feature of heterochromatin. However, many eukaryotes exhibit an additional heterochromatic modification, namely 5-methylcytosine, which is a modification of the DNA itself (Figure 6.4A). While this modification is absent or largely

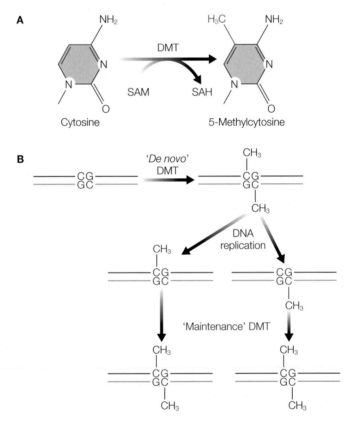

Figure 6.4 *DNA methylation.* (A) The methylation reaction. DNA methyltransferases (DMTs) catalyze the addition of a methyl group to the 5 position of the cytosine ring. The methyl group donor is *S*-adenosylmethionine (SAM). A byproduct of the reaction is *S*-adenosylhomocysteine (SAH). (B) *"De novo"* (meaning "from the new") DMTs select cytosine residues in the DNA for methylation. The selection mechanism is not well understood. Many methylation sites reside in the sequence CG, which can be symmetrically methylated on both strands. After DNA replication, each daughter DNA molecule has inherited one methylated strand and one unmethylated strand. "Maintenance" DMTs recognize the hemimethylated sites and place a methyl group on the newly synthesized strand.

absent from yeast, round worms, and insects, it is an important feature of heterochromatin in higher plants, vertebrates, and filamentous fungi.

Cytosine methylation occurs postreplicatively and eukaryotes contain multiple DNA methyltransferases that catalyze cytosine methylation (Figure 6.4B). As a covalent modification of the DNA itself, cytosine methylation is well suited to epigenetic inheritance. DNA methylation sites are often symmetric having the sequence

$$\begin{array}{cc} \text{CG} & \text{CN G} \\ \text{GC} & \text{or} & \text{GN'C} \end{array}$$

– such sites usually become methylated on both strands. During semiconservative DNA replication, unmethylated C residues are inserted opposite the G residues and thus the resulting double-stranded DNA molecules are "hemimethylated" meaning

> ## What is the mechanistic connection between cytosine methylation and heterochromatic silencing?
>
> Some organisms (e.g., *Drosophila* and yeast) exhibit little or no cytosine methylation, suggesting that it is not essential for silencing. In organisms (e.g., vertebrates, plants) that do exhibit cytosine methylation, the methylation of cytosines and the methylation of histone H3 lysine 9 in heterochromatic regions seem to mutually stimulate one another. Thus, cytosine methylation may help to further reinforce and stabilize the silent state especially in organisms with longer cell cycle and generation times.

that they are methylated on the parental strands, but not on the newly synthesized strands. So-called "maintenance" cytosine methylases recognize the hemimethylated sites and methylate the newly synthesized strand thereby ensuring the reliable inheritance of cytosine methylation.

6.2.6 A distinct mechanism for heterochromatic silencing in budding yeast

One organism in which the histone H3 lysine 9 methylation-dependent pathway for heterochromatin formation is *not* conserved is the budding yeast *Saccharomyces cerevisiae*, which has neither a Clr4 nor a Swi6 ortholog. The complete absence of histone H3 lysine 9 methylation from this genus is surprising, as this modification is present in organisms as diverse as fission yeast, filamentous fungi, protozoans, roundworms, insects, mammals, and flowering plants. Remarkably, however, budding yeast has evolved a different pathway for generating transcriptionally silent heterochromatin.

Heterochromatic silencing in *S. cerevisiae* requires a protein complex termed the SIR (for "silent information regulator") complex, consisting of the Sir2, Sir3, and Sir4 proteins. In many ways this complex functions analogously to Swi6 (although it exhibits no sequence homology to Swi6). It is recruited to DNA by sequence-specific DNA-binding proteins and spreads along chromatin to direct the formation of large transcriptionally silent heterochromatic domains.

Although Sir3 and Sir4 are unique to budding yeast, Sir2, a histone deacetylase that uses NAD^+ as a cofactor, is found throughout the eukaryotic domain. Studies in a number of organisms including fission yeast (see above) and *Drosophila* suggest widespread roles for Sir2 in heterochromatic silencing. Thus, histone deacetylation may be more fundamental for heterochromatic silencing than histone methylation. This could reflect a critical role for histone deacetylation in driving chromatin compaction.

6.3 EPIGENETIC CONTROL BY POLYCOMB AND TRITHORAX GROUP PROTEINS

The vast majority of active genes are found in euchromatin. It is therefore conceivable that heterochromatic silencing is primarily a byproduct of the need to package centromeres and telomeres into a form compatible with reliable chromosome maintenance. If this is true, then the mechanisms of heterochromatic silencing may not be relevant to the epigenetic control of gene activity during development. This view is contradicted, however, by numerous studies of epigenetic transcriptional control suggesting that the mechanisms of heterochromatic silencing *are* used in the developmental control of gene activity. For example, studies examining the mechanisms that regulate homeotic gene expression during embryogenesis hint strongly at the generality of the mechanisms defined for heterochromatic silencing.

6.3.1 Combinatorial control of segment identity

Homeotic genes were first identified and characterized by E.B. Lewis, T.C. Kaufman, W.J. Gehring, and others in studies of *Drosophila* development. These studies defined a set of genes that specify segment identity during development. The body plan of the fly consists of a series of head, thoracic, and abdominal segments, each of which exhibit distinct morphological characteristics. For example, the thorax consists of three segments, T1, T2, and T3, each of which is distinct from the other two (Figure 6.5). Each thoracic segment contains a pair of legs. However, T2 differs from T1 and T3 by the presence of a pair of wings, while T3 differs from T1 and T2 by the presence of a pair of halteres (club-like appendages used for balance during flight).

Mutations in homeotic genes result in homeosis: that is, the assumption by one segment of structural features normally associated with another segment. This is beautifully illustrated by a well-studied mutation of the *Ultrabithorax* (*Ubx*) homeotic gene that results in a fly with two pairs of wings instead of the single pair normally found in flies (Figure 6.5C). Close inspection of the mutant flies reveals that rather than one wing-bearing segment (T2) and one haltere-bearing segment (T3), they contain two

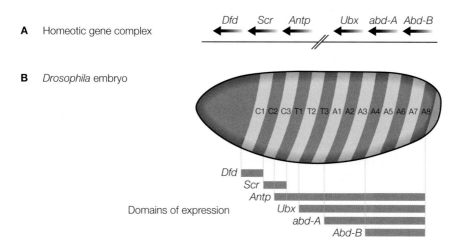

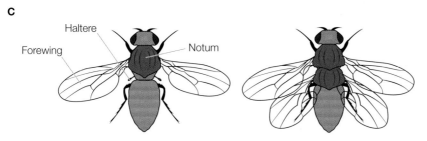

Figure 6.5 *Combinatorial control of segment identity by the homeotic genes.*
(A) The homeotic gene complex in flies contains the six genes shown,
each of which encodes a homeodomain-containing transcription factor.
(B) The *Drosophila* embryo contains 14 domains that will give rise to 14
segments during development. These include the three cephalic segments
(C1–C3), the three thoracic segments (T1–T3), and the eight abdominal
segments (A1–A8). The homeotic genes are expressed in broad
overlapping domains as indicated by the bars underneath the drawing of
the embryo. (C) Mutations in homeotic genes alter segment identity. In
the wild-type fly on the left the T2 segment (in which *Antp* is the only
homeotic gene being expressed) grows a pair of wings, while the T3
segment (in which both Ultrabithorax (Ubx) and Antennapedia (Antp) are
being expressed) grows a pair of halteres. These club-like appendages are
used for balance during flight. The fly on the right contains mutations that
block expression of Ubx in the region of the embryo that will give rise to
T3. As a result, both the T3 and T2 regions in the embryo now lack Ubx
and express only Antp. As a result, they both develop like T2. In other
words, they both give rise to a pair of wings.

wing-bearing segments due to the transformation of T3 into a second copy of T2.

The homeotic genes (including *Dfd*, *Scr*, *Antp*, *Ubx*, *abd-A*, and *Abd-B*) cluster together on a single chromosome (Figure 6.5A), and each one encodes a sequence-specific transcription factor containing a homeodomain DNA-binding motif. These genes are expressed in broad overlapping domains in the early embryo so that different regions of the embryo contain different combinations of the encoded transcription factors (Figure 6.5B).

The determination of segment identity by the homeotic transcription factors is a combinatorial process. While the subject of combinatorial control will be formally introduced in Chapter 7, it is sufficient to explain for now that combinatorial control is what happens when multiple regulatory transcription factors work together to dictate a regulatory outcome. In the case at hand, it is the particular combination of homeotic transcription factors present in any segment that determines how that segment will develop (see legend to Figure 6.5).

6.3.2 Establishment and maintenance phases of homeotic gene expression

Homeotic gene expression can be divided into two phases: the establishment phase which occupies the first half day of embryogenesis and the maintenance phase, which occupies the remaining ~10 days of embryonic, larval, and pupal development culminating in the emergence of the adult fly from the pupal case. The initial establishment of homeotic gene expression domains depends upon conventional sequence-specific transcription factors acting as activators and repressors. The factors responsible for this spatially regulated expression are encoded by segmentation genes, so-called because they direct the subdivision of the embryo into a series of repeating segments. These segmentation transcription factors are well suited to the job of directing the spatially restricted expression of homeotic genes because they are themselves expressed in spatially restricted patterns (see Box 7.2).

The regulation of Ubx by Hunchback provides a particularly nice example of the regulation of a homeotic gene by a segmentation factor (Figure 6.6). Hunchback is only present in the anterior half of the embryo where it functions as a repressor of Ubx. As a result, Ubx expression is restricted to the posterior half of the early embryo.

The factors such as Hunchback that establish spatially regulated homeotic gene expression disappear from the embryo within a few hours after development begins. In contrast, the spatially restricted expression of the homeotic genes is required throughout embryonic development and once again during pupal development (metamorphosis). This dichotomy suggests that the homeotic gene transcriptional states established within the first few hours of development are somehow stabilized before the

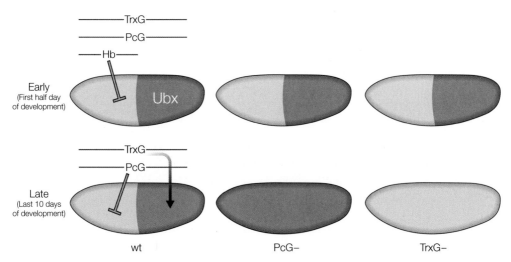

Figure 6.6 *Different mechanisms for the establishment and maintenance of homeotic gene expression.* Initial establishment of the expression domains of the homeotic genes is under the control of sequence-specific transcription factors such as Hunchback (Hb). In the early embryo, Hb, which is only present in the anterior (left) half of the embryo represses *Ubx* transcription and therefore restricts *Ubx* expression to the posterior (right) half of the embryo. Later on in development, after Hb is gone, the maintenance of the proper pattern of *Ubx* expression is dependent on the Trithorax group (TrxG) and Polycomb group (PcG) proteins. These proteins are uniformly distributed throughout the embryo. The TrxG proteins sense when a gene is in the active state and then serve to maintain that active state over multiple cell generations. The PcG proteins sense when a gene is in the silent state and then serve to maintain that silent state over multiple cell generations. As a result, mutations in TrxG genes (TrxG⁻) result in the loss of Ubx activity throughout the late embryo, while mutations in PcG genes (PcG⁻) result in the uniform activation of Ubx throughout the late embryo.

spatially restricted factors such as Hunchback disappear from the embryo. To put it another way, the transcriptional states established in the first several hours of embryogenesis are somehow converted into epigenetic states that can be maintained throughout development.

This epigenetic maintenance phase of homeotic gene expression requires numerous protein factors, most of which can be grouped into one of two sets termed the Polycomb group (PcG) and Trithorax group (TrxG) proteins (Table 6.1). PcG proteins are required to keep genes in the off state – when one or more of the PcG proteins are absent from the embryo all the homeotic genes including *Ubx* are inappropriately activated throughout the embryo after the segmentation factors such as Hunchback disappear from the embryo. Conversely, TrxG proteins are required to keep genes in the on state, and, in their absence, the homeotic genes all fall silent after the disappearance of the segmentation factors (Figure 6.6).

Table 6.1
A subset of Polycomb and Trithorax group proteins

Name	Function and domains
Polycomb group	
PRC1	Recognizes histone H3 trimethyl-lysine 27
Pc	Contains chromodomain
Ph	
Psc	
RING	
PRC2	Trimethylates histone H3 lysine 27
E(z)	Contains SET domain
Esc	
Su(z)12	
Scm	Essential for PcG activity, but function not known
Pho	Contains zinc finger domain, binds PREs and recruits PRC2
Trithorax group	
Trx	Contains SET domain, trimethylates histone H3 lysine 4
Ash1	Contains SET domain, trimethylates histone H3 lysine 4
Brahma complex	ATP-dependent chromatin remodeling complex
Brahma	Homologous to yeast Snf2 – catalytic subunit of the complex
Osa	
Moira	
Questions	

6.3.3 Parallels between heterochromatic and PcG silencing

Lysine methylation followed by chromodomain recognition

To reiterate what was said at the end of the previous section, the PcG proteins somehow sense when a homeotic gene is transcriptionally inactive during early embryogenesis. In response, they direct formation of a silent epigenetic state that it maintained throughout development. The PcG consists of about 15 gene products, many of which function as components of two protein complexes termed Polycomb repressive complex 1 (PRC1) and Polycomb repressive complex 2 (PRC2) (Table 6.1). As we will see, the role of PRC2 in PcG silencing is analogous to the role of Clr4 in heterochromatic silencing, while the role of PRC1 in PcG silencing is analogous to the role of Swi6 in heterochromatic silencing.

The E(z) protein, which is a component of PRC2, contains a SET domain, and, as a result, PRC2 is a histone methyltransferase. In fact, trimethylation of histone H3 lysine 27 by PRC2 is essential for silencing by the PcG. Thus, while Clr4 catalyzed histone H3 lysine 9 di- and trimethylation is an epigenetic modification leading to heterochromatic silencing, PRC2 catalyzed histone H3 lysine 27 trimethylation is an epigenetic modification leading to PcG silencing.

While PRC2 may be the PcG analog of Clr4, PRC1 may be the PcG analog of Swi6. As was discussed in the section on heterochromatic silencing, the di- and trimethylated forms of histone H3 lysine 9 (a hallmark of heterochromatin) bind with high affinity to the chromodomain in Swi6, which then organizes chromatin into a transcriptionally silent state. PRC1 also contains a chromodomain, which, in this case, is highly specific for the trimethylated form of histone H3 lysine 27. Thus lysine 27 trimethylation by PRC2 leads to recruitment of PRC1.

Just as we do not understand how Swi6 recruitment leads to heterochromatic silencing, it is not clear how PRC1 recruitment leads to the silencing of homeotic genes. However, studies using short fragments of chromatin reconstituted *in vitro* suggest that PRC1 may mediate nucleosome aggregation. This aggregation could, in turn, render the template inaccessible to components of the transcriptional apparatus required for gene activity.

Targeting of the PcG

In the case of heterochromatic silencing, we saw that targeting of the silencing machinery involved two mechanisms, one dependent on the RNAi pathway and the other dependent on sequence-specific DNA-binding proteins. In the case of PcG silencing, a role for RNAi in the targeting of PcG complexes has not been established. In contrast, sequence-specific DNA-binding proteins clearly have a role in this process. The homeotic gene complex contains multiple cis-regulatory modules termed Polycomb response elements (PREs). Certain PcG proteins including the zinc finger protein Pho bind specific sites in these PREs. Pho bound to the PREs is thought to recruit PRC2 leading to histone H3 lysine 27 trimethylation, PRC1 recruitment, and silencing (Figure 6.7).

In summary there are multiple parallels between heterochromatic and PcG silencing. Both pathways require the recruitment of a histone H3-specific lysine methyltransferase – Clr4/Su(var)3-9 in the case of heterochromatic silencing and PRC2 in the case of PcG silencing. Lysine di- and/or trimethylation then leads to the recruitment of a chromodomain-containing factor – Swi6 (or its *Drosophila* and human counterpart HP1) in the case of heterchromatic silencing and PRC1 in the case of PcG silencing. The recruitment of this factor may, in each case, render genes inaccessible to the transcriptional machinery.

6.3.4 Maintenance of the active state by TrxG

While PcG proteins maintain the silent state of homeotic genes, TrxG proteins maintain the active state. More specifically, TrxG proteins sense when

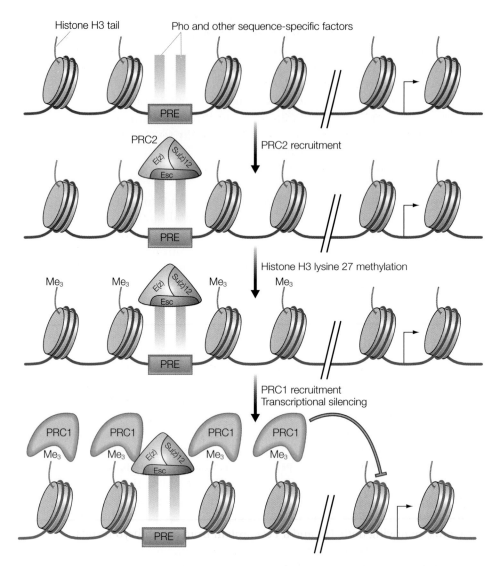

Figure 6.7 *Silencing by the Polycomb group.* Genes that are subject to Polycomb group (PcG) regulation contain cis-regulatory modules termed Polycomb response elements (PREs). These can be located many tens of thousands of basepairs from the promoters being regulated. The PREs contain binding sites for a number of sequence-specific PcG proteins including Pho. These factors recruit PRC2 to the PRE. The SET domain in E(z), a PRC2 subunit, then mediates trimethylation of histone H3 lysine 27. The domain of methylation seems to be limited in size to a few hundred basepairs. This methylated form of histone H3 serves as a docking site for PRC1, since the chromodomain in Pc (a PRC1 subunit) binds to histone H3 trimethyl-lysine 27 with high specificity. PRC1 then silences the promoter. It is not clear how PRC1 is able to act over great distances.

a homeotic gene is transcriptionally active during early embryogenesis. In response, they direct formation of an active epigenetic state and maintain that state throughout development. Like the PcG, the TrxG is a large and complex set of gene products. For the sake of simplicity, this discussion will focus on two subgroups of the TrxG (Table 6.1). The first subgroup consists of Trx and Ash1, two SET domain-containing histone lysine methyltransferases (HKMTs). The second subgroup consists of Brahma, an ATP-dependent chromatin remodeling factor, and associated proteins.

Inhibition of PcG silencing by histone H3 lysine 4 trimethylation

Thus far, the discussion in this chapter has focused on the role of histone methylation in silencing. But as has been mentioned previously (Chapter 5), histone methylation also has roles in activation. While methylation of histone H3 on lysines 9 and 27 is most often associated with silencing, methylation of histone H3 on lysines 4, 36, and 79 is most often associated with gene activation. The two TrxG proteins, Trx and Ash1, catalyze trimethylation of lysine 4 and therefore contribute to the formation of an active epigenetic state.

Genetic studies suggest that lysine 4 trimethylation could lead to maintenance of the active state by antagonizing the function of PcG complexes. As mentioned earlier, mutations in PcG genes result in broad derepression of homeotic gene expression throughout the late embryo, while mutations in TrxG genes result in almost complete loss of homeotic gene expression in the late embryo (Figure 6.6). But what phenotype results when an embryo is mutant for both a PcG gene and a TrxG gene? Is the resulting phenotype similar to one of the single mutant phenotypes or is it intermediate between the two single mutant phenotypes? This question has been addressed by generating embryos mutant for both *trx* and one of the PcG genes. Individuals of this genotype display broad derepression of homeotic gene expression throughout the late embryo (Figure 6.8A). In other words, removing both Trx and a PcG protein from the embryo results in a phenotype that is nearly identical to the PcG single mutant phenotype and the opposite of the *trx* single mutant phenotype.

These double mutant studies indicate that the Trx protein is without a biological function in the absence of an essential component of the PcG. Therefore, the role of Trx (and thus of the histone H3 lysine 4 trimethyl mark) could be to block repression by the PcG (Figure 6.8B). In the absence of a functional PcG, Trx has lost its regulatory target and so taking away Trx in addition to the PcG has no additional consequence.

ATP-dependent chromatin remodeling by Brahma

The TrxG gene *Brahma* encodes the *Drosophila* ortholog of Snf2, which as we saw previously (Chapter 5) is an ATP-dependent chromatin

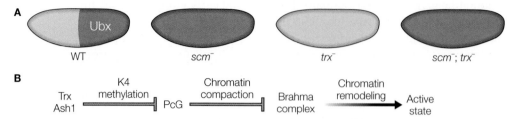

Figure 6.8 *Genetic analysis of the pathway controlling the maintenance of homeotic gene expression.* (A) Phenotypic analysis looking at TrxG;PcG double mutants. A mutation in a Polycomb group (PcG) gene such as *scm* results in loss of repression of homeotic genes such as *Ubx* in the late embryo, while a mutation in a TrxG gene such as *trx* results in the loss of homeotic gene activity in the late embryo. In *scm;trx* double mutants, one observes the same loss of repression phenotype that is seen in *scm* single mutants. Thus, Trx has no function in the absence of Scm suggesting that it works before Scm in the regulatory pathway controlling the maintenance of homeotic gene expression. (B) A possible pathway for the regulation of homeotic gene expression. In accord with the genetic analysis shown in (A) the histone H3 lysine 4 methyltransferases Trx and Ash1 are theorized to inhibit PcG function. Biochemical studies of the PcG complex PRC1 show that it can block remodeling by the Brahma ATP-dependent chromatin remodeling complex, perhaps by mediating the folding of chromatin into a compact and therefore inaccessible state. Thus, the Brahma complex is placed downstream of the PcG in the pathway.

remodeling factor that couples ATP hydrolysis to nucleosome movement ("remodeling"). The Brahma protein associates with a number of additional proteins including the products of several other TrxG genes to form the functional equivalent of the yeast SWI/SNF complex. Such complexes are thought to mediate gene activation by moving nucleosomes out of the way, thereby allowing the transcriptional machinery access to a transcription unit. How does the Brahma complex fit into the overall pathway leading to the stabilization of transcriptional states by PcG and TrxG proteins? Although there is no definitive answer to this question, PRC1 is known to inhibit *in vitro* chromatin remodeling by the Brahma complex. This may reflect the ability of PRC1 to direct the compaction of chromatin, thereby rendering it resistant to remodeling.

6.3.5 A model for the epigenetic regulation of homeotic gene activity

Synthesizing the data on PcG and TrxG proteins and their relatives in other organisms, a simple model to explain how these factors work is as follows (Figure 6.8B):

1 Early during embryogenesis, segmentation genes direct the spatially regulated activation and repression of the homeotic genes. Active

transcription may then be a cue for the recruitment of histone H3 lysine 4-specific methyltransferases such as Trx and Ash1. Trx and Ash1 may then trimethylate histone H3 lysine 4 rendering the chromatin resistant to repression by the PcG.

2 In the absence of homeotic gene transcription during early embryogenesis, histone H3 lysine 4 remains unmethylated allowing the PcG complexes PRC1 and PRC2 to organize the chromatin into a transcriptionally silent state that is resistant to remodeling by Brahma.

Thus, although it is almost certainly an oversimplification, it may be useful to think of the maintenance of homeotic gene activity in terms of a simple linear pathway: the Trx family HKMTs inhibit PcG complexes, which in turn inhibit chromatin remodeling by Brahma.

> **What are the mechanisms responsible for recognizing the transcriptional state of a homeotic gene as established in the early embryo and converting it into a stable epigenetic state?**
>
> Unfortunately, we cannot currently answer this key question with anything other than speculation. One feature that distinguishes actively transcribed genes from silent genes is the presence of Pol II ternary elongation complexes along the length of the transcription unit. Perhaps TrxG proteins are recruited to active genes by some feature of this ternary elongation complex (e.g., the nascent transcript) thereby explaining how the TrxG can "capture" the active state.

6.4 X CHROMOSOME INACTIVATION: PARALLELS TO HETEROCHROMATIC AND POLYCOMB GROUP SILENCING

In the well-known sex determination system used by mammals, females have two X chromosomes per cell, while males have one X chromosome and one Y chromosome per cell. Most of the genes on the X chromosome are not involved in sex-specific functions and thus the products of these genes are required in equal levels in males and females. It is therefore essential for survival of the organism to adjust the level of X-linked gene expression to account for the different X chromosome dosage in males and females. This process is termed dosage compensation.

6.4.1 Random inactivation of the X chromosome

In mammals, dosage compensation occurs by a process of X inactivation. Through a mechanism that is poorly understood, cells "count" the number of X chromosomes per cell. Then, all but one X chromosome in each cell is converted into transcriptionally inactive facultative heterochromatin. As a result, one X chromosome in each normal female cell is inactive. This inactive chromosome (termed Xi) has all the hallmarks of heterochromatin. It is highly condensed and in cells stained with DNA-binding dyes it is visible as a densely staining nuclear body termed the "**Barr body**". As with other heterochromatin, the Barr body is often found just beneath the nuclear membrane and replicates late during the S phase.

In contrast to Xi, the other X chromosome (termed Xa) remains mostly euchromatic and therefore active. This effectively equalizes the X chromosome dosage between males and females since the single X chromosome in each male nucleus also remains active. Note that a consequence of this counting mechanism is that in rare individuals with three X chromosomes per cell (triple-X syndrome) each nucleus contains two Barr bodies and a single active X. Because the extra X chromosome is inactivated in this way, triple-X females have relatively minor developmental abnormalities in contrast to the lethality or much more severe developmental abnormalities associated with extra copies of chromosomes other than the X chromosome.

The selection of the X chromosome to be inactivated in each cell is a random process. Once an X chromosome in an early embryonic cell has been inactivated, that same X chromosome remains inactive through subsequent mitotic cycles. Thus, the active and inactive states of the X chromosome are epigenetic states in the same sense that the transcriptional states of the yeast silent mating type loci and the *Drosophila* homeotic genes are epigenetic states.

Because the X chromosome to be inactivated is selected randomly during early embryogenesis, each tissue in a female mammal is typically a mixture of two phenotypically distinct types of cell clones, a phenomenon that is illustrated by the calico cat. One of the genes controlling coat color in cats, termed the *orange* gene, is on the X chromosome. One allelic form of this gene results in black fur, while another allelic form results in orange fur. In females that are heterozygous for these two alleles, the characteristic calico coat color pattern, containing large patches of orange and black fur, is a result of the random inactivation of the X chromosome early in development (Plate 6.2). Note that with the rare exception of the XXY male (Kleinfelter's syndrome), all calico cats are necessarily female.

6.4.2 Cis-acting RNA in X inactivation

X inactivation in mammals depends on a region on the X chromosome known as the X inactivation center (XIC) (Figure 6.9). This region contains the Xist gene, the RNA products of which do not encode any protein. The process whereby one of the X chromosomes is inactivated begins when one of the two X chromosomes (i.e., the one destined to remain active) is chosen at random to begin transcription of the Xist gene in the antisense direction. The resulting transcript (termed Tsix, which is Xist backwards) then inhibits the synthesis of sense Xist transcripts. This inhibition is somehow restricted to the chromosome from which Tsix is being made; as a result Xist can only be synthesized on the other chromosome (i.e., the one destined to be inactivated). The Xist RNA, which does not

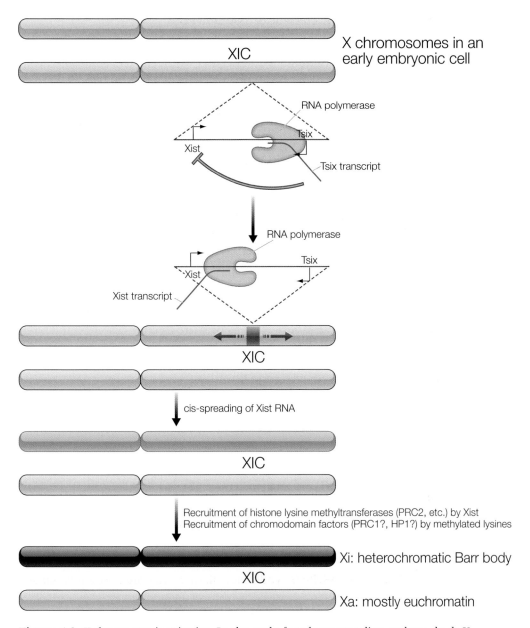

Figure 6.9 *X chromosome inactivation.* In the early female mammalian embryo, both X chromosomes in each cell are euchromatic and therefore active. A locus on the X chromosome termed the X inactivation center (XIC) contains two promoters that direct transcription of the same region in opposite directions. The resulting sense and antisense transcripts are called Xist and Tsix, respectively. By mechanisms that are not understood, a single X chromosome in each cell is randomly selected to initiate transcription from the Tsix promoter. The resulting Tsix transcript then works in cis to block production of the Xist transcript. Thus, Xist transcription is restricted to the other X chromosome. The Xist transcript spreads along the X chromosome from which it is produced and recruits histone lysine methyltransferases such as PRC2. Methylation of this X chromosome then leads to recruitment of additional silencing factors possibly including PRC1 and HP1, leading to heterochromatin formation and X inactivation. The inactive X chromosome (Xi) appears as a compact densely staining body (the Barr body) in female nuclei. The other X chromosome (Xa) remains mostly euchromatic and therefore active.

encode any protein, remains associated with the chromosome from which it arises and spreads from the XIC to coat and inactivate the entire chromosome.

The role of non-coding RNA in X inactivation is at least superficially reminiscent of the role of non-coding RNA (siRNA) in constitutive heterochromatin formation. There are, however, many differences between the two processes. Perhaps most significantly, there is currently no evidence that siRNA spreads along the chromosome in cis from the site at which it is produced. Only time will tell if the involvement of non-coding RNA in X inactivation and constitutive heterochromatin formation is indicative of a deeper mechanistic relationship between these two processes.

6.4.3 Histone modifications characteristic of both heterochromatic and PcG silencing on Xi

The mechanisms by which Xist leads to silencing are beginning to be elucidated. Specifically, Xist RNA appears to recruit histone modifying activities to Xi. These include histone H3 lysine 9 and histone H3 lysine 27 methyltransferases. Thus, the Xi exhibits hallmarks of both heterochromatic and PcG silencing.

The role of histone H3 lysine 27 trimethylation in X chromosome inactivation has been especially well studied. This modification on Xi requires the mammalian counterpart of PRC2, the same lysine 27 methyltransferase required for PcG silencing in *Drosophila*. Expression of Xist RNA is both necessary and sufficient for the transient recruitment of PRC2 leading to histone H3 lysine 27 trimethylation. This transient association occurs at about the time that X chromosome inactivation occurs in the early embryo.

The observation that PRC2 is recruited to Xi at the time of chromosome inactivation strongly suggests a mechanistic connection between X chromosome inactivation and PcG silencing. Since Xi shows all the classical hallmarks of heterochromatin, this provides strong evidence that PcG-silenced genes may be a form of facultative heterochromatin. Thus, the regulatory mechanisms responsible for the silent state of heterochromatin may be applicable to a wide variety of epigenetic phenomena.

6.5 SUMMARY

Transcriptional states can be passed on from cell generation to cell generation during the life of an organism, and may even be transmitted from parent to offspring. Thus, phenotype is strongly influenced by epigenetics in addition to genetics and the environment.

Heterochromatic silencing is a form of epigenetic regulation associated with numerous chromosomal domains including centromeres, telomeres, the yeast silent mating type loci, and the mammalian inactive X chromosome. Studies in the fission yeast _S. pombe_ show that heterochromatic silencing depends on a number of transacting factors including Clr4, a histone H3 lysine 9 methyltransferase, and Swi6, a chromodomain factor that binds histone H3 lysine 9 after its di- and trimethylation by Clr4. By mechanisms that are not well understood, Swi6 then organizes chromatin into condensed domains that are inaccessible to the transcriptional machinery.

Initiation of heterochromatin formation occurs by both RNA-dependent and RNA-independent mechanisms. In RNA-dependent heterochromatin formation, double-stranded RNA resulting from transcription of repeat sequences is processed into siRNA, which then guides a silencing complex termed RITS to the regions of the genome from which the siRNA arose. RITS then directs other components of the silencing machinery (e.g., Clr4, Swi6) to these regions leading to heterochromatin formation.

The mechanisms of heterochromatic silencing are conserved throughout most of the eukaryotic domain. However, budding yeast are an exception to this rule as they lack the histone H3 lysine 9 methylation-dependent pathway for silencing. Instead, budding yeast have evolved a different system utilizing the products of the SIR genes for the formation of heterochromatin.

Another type of epigenetic control involves the action of the Polycomb group (PcG) and Trithorax group (TrxG) proteins. In _Drosophila_, these groups of proteins are critical for the maintenance of homeotic gene transcriptional states throughout the developmental cycle of the organism. Spatially regulated transcription of the homeotic genes is established through the action of sequence-specific transcription factors encoded by the segmentation genes. However, these transcription factors disappear within a half-day after development begins and it is the job of PcG proteins to maintain the off state and of TrxG proteins to maintain the on state of homeotic genes for the remainder of the developmental cycle.

PcG proteins maintain genes in the off state by mechanisms that are remarkably parallel to those used in heterochromatic silencing. Like heterochromatic silencing, PcG-dependent silencing depends on a histone lysine methyltransferase. In this case, the methyltransferase, which is termed PRC2, catalyzes histone H3 lysine 27 trimethylation. The trimethylated residue is recognized by a chromodomain-containing complex termed PRC1, which then triggers the formation of the silent state.

The maintenance of the active state by TrxG proteins may involve antagonistic interactions between TrxG and PcG proteins. For example, methylation of histone H3 lysine 4 by the TrxG methyltransferases Trx and Ash1 may block repression by PcG proteins. PcG proteins may in turn block chromatin remodeling by the TrxG protein Brahma, which is the catalytic subunit of an ATP-dependent chromatin remodeling complex.

The inactive X chromosome found in the cells of female mammals exhibits hallmarks of both heterochromatic silencing and PcG-dependent silencing suggesting a unity between these two types of epigenetic regulation. The inactivation of all but one X chromosome in each mammalian cell depends on a cis-acting non-coding transcript termed Xist, which spreads along the X chromosome and attracts histone modifying activities to the chromosome.

Like other forms of heterochromatin, the inactive X is highly condensed, resides at the nuclear periphery, and is enriched in histone H3 dimethylated at lysine 9. Like other PcG-silenced loci, the inactive X is enriched in histone H3 trimethylated at lysine 27. Furthermore, this latter form of methylation requires the mammalian counterpart of PRC2, which associates with the X chromosome during early embryogenesis just as it is being inactivated.

PROBLEMS

1 What phenotype do you predict for a *brahma*, *pc* double mutant fruit fly embryo. In other words, how would the phenotype compare to that of *brahma* and *pc* single mutants?

2 The PREs in the *Drosophila* homeotic gene complex are closely associated with TREs (Trithorax response elements). TREs are often transcribed, giving rise to non-coding transcripts. A paper published in *Science* (Sanchez-Elsner, T., Gou, D., Kremmer, E. and Sauer, F. (2006) Noncoding RNAs of trithorax response elements recruit Drosophila Ash1 to Ultrabithorax. *Science*, **311**(5764), 1118–1123) shows that the transcription of a TRE responsible for regulation of Ubx expression occurs coordinately with the transcription of Ubx (i.e., whenever the *Ubx* gene is transcribed, the TRE is also transcribed). Furthermore, Ash1 appears to bind to the non-coding transcripts, which then recruit Ash1 to the Ubx locus. How might recruitment of Ash1 by TRE transcripts assist in the maintenance of the transcriptionally active state by the Trithorax group?

3 How do you expect mutations in the *argonaut* and *clr4* genes to alter the post-translational modification state of centromeric chromatin in *S. pombe*? Would you expect these mutations to have the same affect on the modification state of the *K*-region? Explain.

4 How does the appearance of the calico cat provide support for the idea that the active and inactive states of the X chromosome are epigenetically stable states?

FURTHER READING

Heterochromatic silencing

Grewal, S.I. and Klar, A.J. (1996) Chromosomal inheritance of epigenetic states in fission yeast during mitosis and meiosis. *Cell*, **86**, 95–101. *Demonstration that epigenetic transcriptional states can obey the laws of Mendelian inheritance.*

Nakayama, J., Rice, J.C., Strahl, B.D., Allis, C.D. and Grewal, S.I. (2001) Role of histone H3 lysine 9 methylation in epigenetic control of heterochromatin assembly. *Science*, **292**, 110–113.

Yamada, T., Fischle, W., Sugiyama, T., Allis, C.D. and Grewal, S.I. (2005) The nucleation and maintenance of heterochromatin by a histone deacetylase in fission yeast. *Mol Cell*, **20**, 173–185. *The roles of histone methylation and deacetylation in the maintenance of the transcriptionally silent heterochromatic state.*

Verdel, A., Jia, S., Gerber, S., Sugiyama, T., Gygi, S., Grewal, S.I. and Moazed, D. (2004) RNAi-mediated targeting of heterochromatin by the RITS complex. *Science*, **303**, 672–676. *The initiation of heterochromatic silencing by double-stranded RNA.*

Jia, S., Noma, K. and Grewal, S.I. (2004) RNAi-independent heterochromatin nucleation by the stress-activated ATF/CREB family proteins. *Science*, **304**, 1971–1976. *The initiation of heterochromatic silencing by sequence-specific transcription factors.*

Schotta, G., Ebert, A., Krauss, V., Fischer, A., Hoffmann, J., Rea, S., Jenuwein, T., Dorn, R. and Reuter, G. (2002) Central role of Drosophila SU(VAR)3-9 in histone H3-K9 methylation and heterochromatic gene silencing. *EMBO J*, **21**, 1121–1131. *Position effect variegation and conserved mechanisms of heterochromatic silencing.*

Liou, G.G., Tanny, J.C., Kruger, R.G., Walz, T. and Moazed, D. (2005) Assembly of the SIR complex and its regulation by O-acetyl-ADP-ribose, a product of NAD-dependent histone deacetylation. *Cell*, **121**, 515–527. *The pathway to heterochromatic silencing in budding yeast.*

DNA methylation and epigenetic inheritance

Saze, H., Mittelsten Scheid, O. and Paszkowski, J. (2003) Maintenance of CpG methylation is essential for epigenetic inheritance during plant gametogenesis. *Nat Genet*, **34**, 65–69. *The essential role of maintenance DNA methyltransferases in epigenetic inheritance.*

Chan, T.L., Yuen, S.T., Kong, C.K., Chan, Y.W., Chan, A.S., Ng, W.F., Tsui, W.Y., Lo, M.W., Tam, W.Y., Li, V.S. and Leung, S.Y. (2006) Heritable germline epimutation of MSH2 in a family with hereditary nonpolyposis colorectal cancer. *Nat Genet*, **38**, 1178–1183. *DNA methylation, epigenetic inheritance, and colon cancer.*

Jacobsen, S.E. and Meyerowitz, E.M. (1997) Hypermethylated SUPERMAN epigenetic alleles in arabidopsis. *Science*, **277**, 1100–1103. *DNA methylation, epigenetic inheritance, and flower development.*

Epigenetic control by Polycomb and Trithorax group proteins

Zhang, C.C. and Bienz, M. (1992) Segmental determination in Drosophila conferred by hunchback (hb), a repressor of the homeotic gene Ultrabithorax (Ubx). *Proc Natl Acad Sci USA*, **89**, 7511–7515. *Demonstration that the establishment of Ubx expression relies on a sequence-specific factor (Hunchback) while the maintenance of Ubx expression relies on the Polycomb group.*

Cao, R., Wang, L., Wang, H., Xia, L., Erdjument-Bromage, H., Tempst, P., Jones, R.S. and Zhang, Y. (2002) Role of histone H3 lysine 27 methylation in

Polycomb-group silencing. *Science*, **298**, 1039–1043. Wang, L., Brown, J.L., Cao, R., Zhang, Y., Kassis, J.A. and Jones, R.S. (2004) Hierarchical recruitment of polycomb group silencing complexes. *Mol Cell*, **14**, 637–646. *Two papers demonstrating the role of histone methylation and chromodomains in Polycomb group silencing.*

Shao, Z., Raible, F., Mollaaghababa, R., Guyon, J.R., Wu, C.T., Bender, W. and Kingston, R.E. (1999) Stabilization of chromatin structure by PRC1, a Polycomb complex. *Cell*, **98**, 37–46. Francis, N.J., Kingston, R.E. and Woodcock, C.L. (2004) Chromatin compaction by a polycomb group protein complex. *Science*, **306**, 1574–1577. *Two papers demonstrating that PRC1-mediated condensation of chromatin may yield a state that is inaccessible to chromatin-remodeling factors.*

Klymenko, T. and Muller, J. (2004) The histone methyltransferases Trithorax and Ash1 prevent transcriptional silencing by Polycomb group proteins. *EMBO Rep*, **5**, 373–377. *Genetic demonstration that histone H3 lysine 4 methylation inhibits Polycomb group function.*

X chromosome inactivation

Clemson, C.M., McNeil, J.A., Willard, H.F. and Lawrence, J.B. (1996) XIST RNA paints the inactive X chromosome at interphase: evidence for a novel RNA involved in nuclear/chromosome structure. *J Cell Biol*, **132**, 259–275. *Study showing that Xist RNA coats the entire inactive X chromosome (the Barr body).*

Plath, K., Fang, J., Mlynarczyk-Evans, S.K., Cao, R., Worringer, K.A., Wang, H., de la Cruz, C.C., Otte, A.P., Panning, B. and Zhang, Y. (2003) Role of histone H3 lysine 27 methylation in X inactivation. *Science*, **300**, 131–135. *Study showing that histone H3 methylation by the mammalian counterpart of PRC2 may be required for X chromosome inactivation.*

7

Combinatorial control in development and signal transduction

Key concepts

- Combinatorial control, in which multiple activators, repressors, and coregulators interact to dictate the transcriptional state of a gene, increases regulatory diversity in development and signal transduction
- Synergy and antagonism represent two major forms of combinatorial control
- Synergy allows the integration of multiple positive inputs into a regulatory decision, and can be mediated by the cooperative assembly of nucleoprotein complexes termed enhanceosomes
- Antagonism allows the integration of both positive and negative inputs into a regulatory decision. It occurs by a variety of mechanisms including competitive protein–DNA and protein–protein interactions

7.1 INTRODUCTION

The unique identity of each tissue in a multicellular organism results from the expression of a unique set of genes. These different patterns of gene activity first arise during development and are further refined in the mature organism in response to external signals. Individual genes are often regulated by multiple cues, including spatial cues, which allow a cell to determine its position within a developing organism and differentiate accordingly, as well as environmental cues, which allow a cell to sense changes in the environment and mount a relevant response. These cues are conveyed by combinations of regulatory transcription factors converging upon the complex cis-regulatory modules (CRMs) associated with each gene. These CRMs integrate many types of information to produce highly refined transcriptional responses. While this type of control, termed combinatorial control, occurs in all organisms, it is especially well developed in multicellular eukaryotes, reflecting the intricately regulated gene expression patterns required by complex organisms.

Five percent or more of the genes in multicellular organisms encode regulatory transcription factors. The *Drosophila* genome, for example, may encode ~700 such factors, while the human genome may encode twice that number. Despite these large numbers, it would be impossible to achieve the intricate patterns of gene regulation required for multicellular life without combinatorial control. Not only does combinatorial control greatly amplify the types of regulation that can be achieved with a given set of factors, it also makes possible certain types of control that would otherwise be impossible. For example, combinatorial control makes it possible to combine overlapping patterns of gene expression to generate novel patterns of gene expression. In addition, combinatorial control makes it possible for tissues to respond in a highly refined manner to environmental signals leading to appropriate adaptive changes in the transcriptional program of the cell.

This chapter will begin with an introduction to two major forms of combinatorial control used by all organisms, namely synergy and antagonism. Synergy will then be illustrated through an introduction to the enhanceosome, with particular emphasis on the well-characterized human interferon-β (IFNβ) enhanceosome. Antagonism will be illustrated through an introduction to the mechanisms that direct the formation of transcriptional stripes during animal development; and through a discussion of signal-mediated switches, in which signals are transduced to the nucleus where they switch genes from a silent to an active state.

7.2 SYNERGY AND ANTAGONISM

7.2.1 Integration of regulatory inputs by cis-regulatory modules

In multicellular eukaryotes, CRMs are composed of multiple binding sites for a variety of sequence-specific activators and repressors, and each gene is typically regulated by multiple CRMs. Combinatorial interactions can occur between multiple factors bound to a single CRM, between factors bound to different CRMs, and between factors bound to a CRM and factors bound at the core promoter. In this context, two factors are said to interact if they collaborate in some way to influence the transcriptional output from a promoter. This does not always involve a direct physical interaction between the two factors.

A direct physical interaction between a factor bound to a distant CRM and a promoter-bound factor would require the formation of a DNA loop to bring the two factors together. While such looping undoubtedly occurs (see Chapter 1), DNA looping is by no means the only mechanism to explain the ability of CRMs to act over long distances. For example, as already discussed in Chapter 6, factors recruited to a CRM in the *Schizosaccharomyces pombe K*-region recruit additional factors leading to the spread along chromatin of a transcriptionally silent domain from the CRM to the promoters of the silent mating-type loci.

7.2.2 The "AND" and "NOT" operators

At the level of the transcriptional output, combinatorial control often boils down to one of two things, synergy or antagonism (Figure 7.1). Synergy occurs whenever the combined effect of two factors is more than the sum of the effects of each factor functioning alone. In the extreme case, significant promoter activity is only observed when both factors are present. In the language of mathematics, this type of combinatorial control can be described by an AND operator, since promoter activity requires the presence of factors X AND Y. This type of combinatorial control provides a way to ensure that genes will only be active in the presence of multiple positive regulatory inputs.

There are several possible mechanistic explanations for synergy. Two activators might facilitate different steps in gene activation (e.g., chromatin remodeling and Pol II recruitment), each of which is critical for the production of a functional transcript. In the absence of either factor, transcription would therefore not occur. Alternatively, multiple activators might bind to DNA cooperatively or cooperatively recruit coactivators to a CRM (Box 7.1). An example of synergy that results from cooperativity will be presented in the upcoming section on enhanceosomes.

While synergy entails interactions between activators, antagonism occurs when repressors interfere with the function of activators. Antagonism can

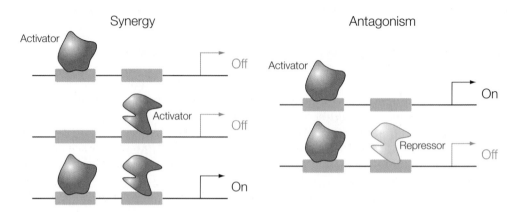

Figure 7.1 *Synergy and antagonism.* In synergy, promoter activity in the presence of two activators is greater than the sum of the promoter activity obtained in the presence of each factor alone. In the extreme case, shown here, promoter activity absolutely requires the presence of multiple factors. Synergy provides a means of ensuring that a gene is only activated when multiple conditions are fulfilled. In antagonism, a repressor counteracts the effect of an activator. In the example shown here, the activator and repressor bind to distinct sites allowing the repressor to block the activator. Alternatively, the activator and repressor can bind to the same site, in which case binding becomes mutually exclusive.

be described by a NOT operator because gene activity requires the presence of an activator ("factor X") and the absence of a repressor ("factor Y"). In other words, gene activity requires fulfillment of the condition "X NOT Y". This type of control provides a means of weighing positive and negative regulatory inputs against one another before deciding whether or not to activate a gene.

Just like synergy, antagonism results from a variety of mechanisms. In long-range repression, antagonism occurs regardless of the relative position of the activator- and repressor-binding sites in the gene. This is exemplified by heterochromatic silencing or Polycomb group silencing (see Chapter 6), two phenomena in which entire loci are organized into conformations that are inaccessible to activators. In short-range repression, antagonism only occurs when repressor- and activator-binding sites are appropriately positioned relative to one another. For example, if an activator and repressor bind to the same or overlapping sites, binding is likely to be mutually exclusive leading to antagonism. An example of this kind of mutually exclusive binding will be presented in the upcoming section on signal-mediated switches.

Short-range repression does not always require overlapping repressor- and activator-binding sites. As will become apparent in the upcoming section on stripe formation, it is sometimes sufficient for the repressor- and activator-binding sites to be located within ~100 bp of one another

Box 7.1

Cooperative binding and synergy

Many examples of synergy, such as the synergy that occurs at the IFNβ enhancer can be explained by the idea of cooperative binding. Cooperativity is a subject that is susceptible to rigorous physical (i.e., mathematical) analysis. The math will not be presented here, however, since, an intuitive appreciation of cooperativity is sufficient for an understanding of most of the combinatorial control literature (including this chapter).

A process is cooperative if it occurs in multiple steps and if the earlier steps make the later steps easier. In a common example often covered in biochemistry courses, the binding of oxygen to hemoglobin is cooperative because the binding of the first molecule of oxygen stabilizes a conformation with increased affinity for additional molecules of oxygen. Similarly, the binding of multiple proteins to a CRM is cooperative if the binding of one protein increases the affinity of the CRM for a second protein.

Different mechanisms can account for cooperative binding of transcription factors. For example, the binding of one protein to the DNA might stabilize a DNA conformation that has increased affinity for a second protein. An example of this type of cooperativity is found at the IFNβ enhanceosome. The NF-κB-binding site in the IFNβ enhancer is intrinsically bent in such a way as to be incompatible with NF-κB binding. However, when the architectural factor HMG I binds to this site, it straightens the DNA and thus favors the binding of NF-κB.

Perhaps a more common type of cooperative binding involves contacts between bound proteins. If activator X makes favorable contact with activator Y when both are bound to sites in a CRM, then the binding of X will favor the binding of Y (Figure B7.1A). This is true because, in the presence of X, the binding of Y is stabilized by favorable interactions with the DNA *and* X. An important feature of cooperativity is that it is mutual – just as X stabilizes the binding of Y, Y stabilizes the binding of X. Consequently, concentrations of Y that are insufficient for high levels of binding to the CRM in the absence of X may be sufficient for high levels of binding in the presence of X. Similarly, concentrations of X that are insufficient for high levels of binding in the absence of Y may be sufficient for high levels of binding in the presence of Y. In cases such as these, X and Y will be observed to function synergistically because the X- and Y-binding sites will only be occupied at high levels when both activators are present.

The idea of cooperativity can be readily extended to situations involving coactivators that simultaneously contact multiple activators. For example, imagine coactivator Z, which simultaneously contacts activators X and Y (Figure B7.1B). When both activators occupy the CRM, the binding of the coactivator will be favored relative to a situation in which one of the activators is absent. Consequently, concentrations of the coactivator that are insufficient for high levels of binding in the absence of either activator may be sufficient for high levels of binding in the presence of both activators. Once again, the activators will be observed to function synergistically, since the coactivator will only bind efficiently in the presence of both activators. This is exactly the situation that occurs at the IFNβ enhancer except that, in this case, three activators, rather than two, simultaneously contact coactivators such as the STAGA complex and CBP.

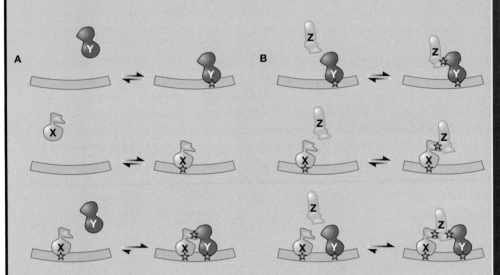

Figure B7.1 *The molecular basis of cooperative binding.* (A) Cooperative binding of two activators to adjacent sites. (B) Cooperative recruitment of a coactivator Z by two adjacently bound activators X and Y. The stars mark energetically favorable binding interactions. The relative favorability of the reactions is indicated by the relative sizes of the forward and reverse arrows.

for the repressor to block activator function. Very little is known about the mechanisms responsible for this kind of short-range repression – we can only speculate that it may involve local changes in chromatin structure or direct interactions between activators and repressors that only occur over short distances along the DNA.

7.3 SYNERGY AND THE ENHANCEOSOME

7.3.1 Enhanceosome assembly and architectural factors

As mentioned above, the term transcriptional synergy describes the phenomenon in which multiple regulatory factors contribute in a greater than additive manner to the activation of a promoter. The so-called "enhanceosome" provides a mechanistic explanation for synergy. This term describes an enhancer (a type of CRM, see Chapter 1) to which multiple activators and coactivators are simultaneously bound, interacting both with one another and the chromatin to form a nucleoprotein complex. Optimal enhanceosome function normally requires the presence of all the components of the complex – the absence of any one component usually leads to enhanceosome destabilization and therefore to a large decrease in transcriptional activation. This is another way of saying that enhanceosome assembly is cooperative, i.e., the binding of one factor increases the affinity of the enhancer for other factors (Box 7.1).

Much of what we know about enhanceosomes derives from studies of the IFNβ gene (Figure 7.2). The expression of IFNβ, a potent antiviral protein, is induced by viral infection, which leads to the release of double-stranded RNA from dying, infected cells. This RNA signals through transmembrane receptors to trigger increases in the activity of multiple sequence-specific transcription factors. These factors include ATF-2/c-Jun, NF-κB, and IRF, all of which bind sites in the IFNβ enhancer. The enhancer then integrates the effects of these multiple activators through the enhanceosome.

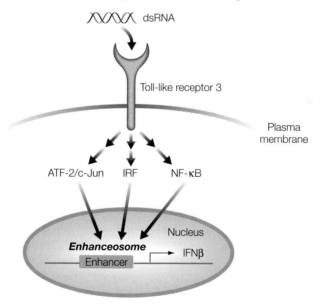

Figure 7.2 *IFNβ activation.* IFNβ is an antiviral protein, the expression of which is induced by the release of double-stranded RNA (dsRNA) from dying infected cells. The dsRNA binds to Toll-like receptor 3 stimulating multiple downstream signal transduction pathways. This leads to the activation of three sequence-specific activator proteins ATF-2/c-Jun, IRF, and NF-κB, which act synergistically to activate IFNβ expression. The three factors assemble with one another at the IFNβ enhancer to form the enhanceosome.

This allows for a much more potent response than would be possible with a single activator.

In the absence of other factors, the binding of the three sequence-specific factors mentioned above to their adjacent recognition sites in the IFNβ enhancer is *not* cooperative. However, the situation changes upon addition of a fourth sequence-specific factor termed HMG I, which binds to multiple sites in the enhancer. In the presence of HMG I, the binding of the sequence-specific factors is rendered cooperative. The capacity of HMG I to exert this effect may reflect its ability to alter the curvature of DNA, thereby allowing factors bound to adjacent sites to make the favorable contacts with one another required for cooperativity. Factors such as HMG I, which facilitate interactions between other factors by changing the curvature of the DNA, are often termed architectural factors.

7.3.2 Cooperative recruitment of coactivators by enhanceosomes

How might formation of an enhanceosome lead to transcriptional activation? Once again, important clues come from studies of the IFNβ enhanceosome. These studies show that the three major sequence-specific activators (ATF-2/c-Jun, NF-κB, and IRF) that bind the enhancer interact with multiple coactivators. The interactions between any one these activators and a coactivator are of generally low affinity. However, when all three activators bind to the enhancer, they combine to form a high affinity interaction surface for coactivator recruitment. In other words, simultaneous contacts between a coactivator and multiple activators cooperatively stabilize the association of the coactivator with the enhanceosome.

Evidence that cooperative coactivator recruitment is required for activation of the IFNβ promoter comes from studies varying the spacing between the transcription factor-binding sites (Figure 7.3). These studies show that increasing the spacing between adjacent binding sites by an integral multiple of the DNA helical repeat length (e.g., ~10 bp) does not interfere with enhancer function, while increasing the spacing by a non-integral multiple of the helical repeat length (e.g., ~5 bp) prevents enhanceosome assembly and consequently prevents activation. This is just as would be expected if alignment of multiple factors on the same face of the DNA were essential to allow cooperative recruitment of coactivators. While increasing the spacing between two binding sites by 10 bp would not disrupt the alignment, increasing the spacing by 5 bp would rotate one binding site to the opposite face of the DNA helix relative to an adjacent binding site thereby preventing activators bound to these sites from simultaneously contacting a coactivator.

> **Helical repeat length** – the number of residues required to complete one turn of a helix. In double-stranded DNA the average helical repeat length is slightly more than 10 bp.

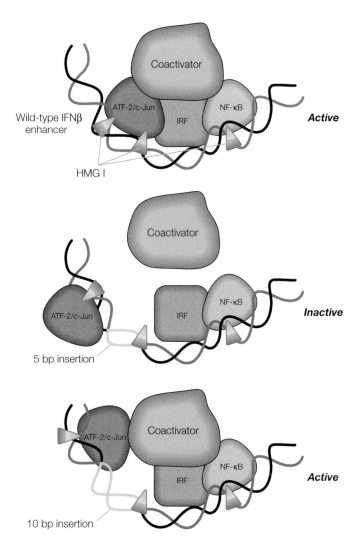

Figure 7.3 *Cooperative enhanceosome assembly.* ATF-2/c-Jun, IRF, and NF-κB bind cooperatively to the IFNβ enhancer. Cooperative binding requires the architectural factor HMG I which binds to three sites in the enhancer and facilitates cooperative interactions by altering the curvature of the DNA. These factors then form a platform for the cooperative recruitment of coactivators. Evidence that cooperative coactivator recruitment is required for enhanceosome function comes from experiments in which inserts of various lengths are introduced between the transcription factor-binding sites. A 5 bp insertion prevents enhanceosome assembly and transcriptional activation, while a 10 bp insertion does not. This is probably because the 5 bp insertion, which corresponds to half a helical turn of DNA, disrupts the alignment between the factors preventing cooperative coactivator recruitment. In contrast, a 10 bp insertion, which corresponds to one helical turn of DNA, would not be expected to disrupt alignment of the factors.

7.3.3 Sequential coactivator recruitment

Enhanceosomes often recruit multiple coactivators in a well-defined temporal sequence. Sequential coactivator recruitment has been nicely illustrated by experiments monitoring the timing of coactivator recruitment by the IFNβ enhanceosome (Figure 7.4). These experiments show that the sequence-specific factors of the enhanceosome begin the job of activating the promoter by recruiting the STAGA histone acetyltransferase (HAT) complex (see Chapter 5). Once recruited to the enhancer, STAGA acetylates the histone H3 and H4 tails leading, by unknown mechanisms, to the ejection of STAGA and the recruitment of a coactivator termed CBP (see Chapter 4) to the enhancer. CBP binds Pol II and thus recruits Pol II to the template. At the same time, the SWI/SNF ATP-dependent chromatin remodeling complex (see Chapter 5) is brought to the promoter, in part through interactions with Pol II and also probably through interactions between the bromodomain in Snf2 (the catalytic subunit of the SWI/SNF complex) and histone acetyl-lysine residues created by the action of the STAGA complex. The SWI/SNF complex is then thought to mobilize nucleosomes or loosen their association with the template, as required for promoter clearance and processive transcriptional elongation.

> **STAGA complex** – the human equivalent of the yeast SAGA histone acetyltransferase complex (see Chapter 5).

The above sequence of events defined for the IFNβ enhanceosome and promoter cannot be generalized to all enhanceosomes. In some cases, for example, SWI/SNF complex recruitment appears to precede recruitment of HAT complexes. What each enhanceosome has in common is the ability to cooperatively recruit multiple coactivators, each of which facilitates a different step in the pathway toward gene activation.

> **What determines the order in which coactivators are recruited to an enhancer?**
>
> As we have seen, enhancers interact with multiple coactivators such as HAT complexes, ATP-dependent chromatin remodeling complexes, and the Mediator. The order in which these factors are recruited to an enhancer appears to differ from one enhancer to the next. We do not understand what determines the order of coactivator recruitment, nor do we understand how changes in the order of coactivator recruitment might change the magnitude or timing of a transcriptional response.

7.4 ANTAGONISM AND STRIPE FORMATION

While synergy (the AND operator) ensures that genes will only be activated in the presence of multiple positive inputs, antagonism (the NOT operator) allows positive and negative inputs to be weighed against one

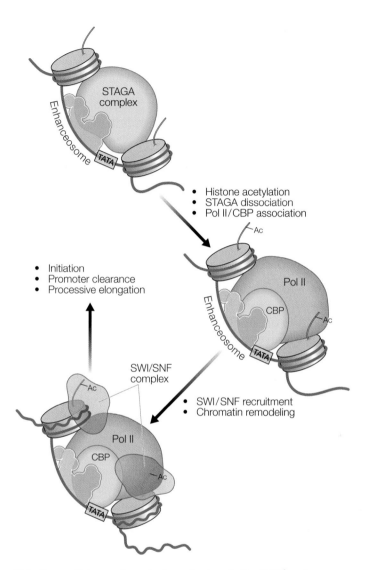

Figure 7.4 *Sequential recruitment of coactivators to the IFNβ enhanceosome.* Activation by the IFNβ enhanceosome is thought to involve sequential recruitment of the STAGA histone acetyltransferase complex followed by recruitment of CBP (CREB-binding protein) and Pol II followed by the recruitment of the SWI/SNF complex. The STAGA complex acetylates flanking nucleosomes. This may trigger STAGA release and CBP recruitment. Recruitment of the SWI/SNF complex may involve contacts between SWI/SNF and Pol II as well as binding of the Snf2 bromodomain to histone acetyl-lysine residues. Chromatin remodeling by SWI/SNF then facilitates promoter clearance and processive elongation. Ac, acetyl group.

another in the decision to activate a gene. As will become apparent in this section, antagonism is used to define the boundaries of a domain of gene activity leading, for example, to stripe formation.

7.4.1 Developmental regulatory networks

Combinatorial control plays many essential roles in the control of transcription during development. This is beautifully exemplified by the regulation of *Drosophila melanogaster* embryogenesis. In the 1980s and 1990s, C. Nusslein-Volhard, E. Weischaus, and others undertook screens to identify many of the genes controlling *Drosophila* embryogenesis. A large fraction of these genes have been found to encode transcription factors that are expressed in spatially restricted domains in the early embryo and direct the subdivision of the organism into multiple segments and tissues giving rise to the complex body plan of the mature embryo (Box 7.2). These transcription factors constitute a regulatory network: the transcription factors encoded by the earliest acting genes control the expression of the next set of genes, which in turn encode transcription factors controlling the next set of genes, and so on. Combinatorial interactions between transcription factors acting at any given step in the hierarchy allow for the gradual refinement of expression patterns, with initial broad expression domains giving rise to more and more refined expression patterns, leading, for example, to narrow stripes of transcription that define the boundaries between segments.

7.4.2 Short-range repression and stripe formation

Making stripes inelegantly

Soon after the onset of *Drosophila* embryogenesis, the gap genes are expressed in broad overlapping domains along the length of the embryo (Box 7.2). These genes encode transcription factors that regulate the expression of the pair-rule genes, so named because mutations in them result in the loss of every other segment from the embryo. Thus, instead of the 14 segments found in the wild-type organism, embryos containing a mutation in one of the pair-rule genes develop only seven segments. In accord with the requirement for each pair-rule gene product in every other segment, they are each expressed in seven regularly spaced stripes along the length of the embryo (Box 7.2).

How might such regularly repeating stripes of gene expression arise? Results from a branch of chemistry known as non-equilibrium thermodynamics show that uniformly distributed factors that control one another's synthesis in just the right way can spontaneously arrange themselves in stripes. Initial attempts to determine the mechanisms of stripe formation were therefore based on the hypothesis that embryos might contain "self-organizing" systems that would spontaneously generate stripes.

Box 7.2

Transcription factor networks and gradients in *Drosophila* segmentation

The gene network that regulates the subdivision of the *Drosophila* embryo into a series of segments provides many beautiful examples of combinatorial control by transcription factors. Cloning and characterization of the genes controlling this process revealed that most of them encode transcription factors that work together in a hierarchy to control one another's expression (Figure B7.2A).

One of the earliest acting genes in this hierarchy is *bicoid*, which encodes a transcription factor containing a homeodomain DNA-binding domain (Figure B7.2B). Studies of this factor were of seminal importance in explaining how different regions within an initially homogeneous embryo can begin to express different genes. The *bicoid* mRNA is made during oogenesis and becomes trapped at one end of the oocyte. After fertilization, while the embryo is still a single cell, translation of this

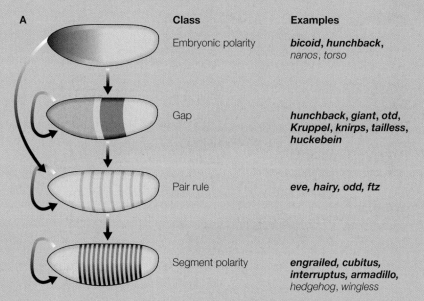

	Class	Examples
A	Embryonic polarity	**bicoid, hunchback,** *nanos, torso*
	Gap	**hunchback, giant, otd, Kruppel, knirps, tailless, huckebein**
	Pair rule	**eve, hairy, odd, ftz**
	Segment polarity	**engrailed, cubitus, interruptus, armadillo,** *hedgehog, wingless*

Figure B7.2 *Transcription factor networks and segmentation.* (A) Segmentation of the embryo is controlled by genes that are expressed in patterns of increasing intricacy as development proceeds. The embryonic polarity genes are expressed in gradients, the gap genes are expressed in broad domains, the pair-rule genes are expressed in seven stripes, and the segment polarity genes are expressed in 14 stripes leading to subdivision of the embryo into 14 segments. A selection of genes in each class is listed with the genes encoding the transcription factors indicated in bold (e.g., **Kruppel**). The arrows indicate regulatory interactions between genes at various levels of the hierarchy.

▶

localized mRNA and the diffusion of the resulting transcription factor away from the site of translation produce a concentration gradient of the Bicoid protein. Bicoid target genes contain CRMs that respond to different concentrations of Bicoid, and are therefore activated in different domains along the length of the embryo. For example, the *hunchback* CRM is sensitive to low concentrations of Bicoid and therefore *hunchback* is activated throughout the anterior half of the embryo. In contrast, the *giant* CRM is only sensitive to high concentrations of Bicoid and therefore *giant* is only activated in the anterior ~30% of the embryo.

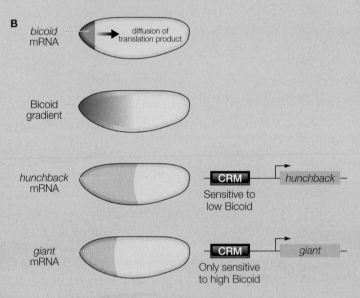

Figure B7.2 *(continued) Transcription factor networks and segmentation.* (B) Differential sensitivity of CRMs to Bicoid leads to different domains of gap gene expression. See text.

However, experimental analysis of pair-rule gene expression revealed that the true mechanism of stripe formation does not involve an elegant self-organizing system. Rather, pair-rule genes were found to contain multiple CRMs (in this case termed "stripe enhancers"), each of which contains the instructions required for the formation of one or two of the seven stripes of pair-rule gene expression.

Stripe formation has been most carefully analyzed in studies of the pair-rule gene *even-skipped* (*eve*), which contains no less than five stripe enhancers, three of which direct the formation of a single stripe each

Non-equilibrium thermodynamics – a branch of chemistry that describes the behavior of systems that are not at equilibrium. Non-equilibrium thermodynamics provides a physical explanation for seemingly impossible events such as the spontaneous reorganization of homogeneous mixtures of chemicals into stripes.

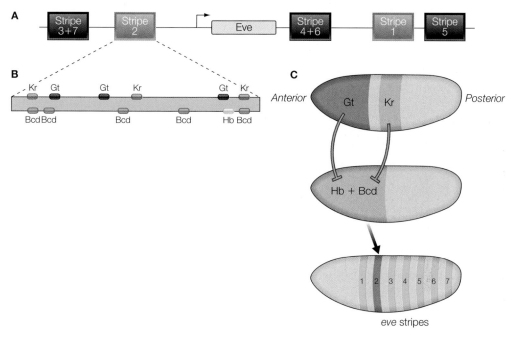

Figure 7.5 *Stripe formation through antagonism.* (A) The *even-skipped* (*eve*) gene contains five stripe enhancers that direct the expression of *eve* in seven stripes. (B) The stripe 2 enhancer contains binding sites for two activators termed Hunchback (Hb) and Bicoid (Bcd) and two repressors termed Giant (Gt) and Kruppel (Kr). (C) Hb and Bcd are present in a broad anterior domain, while Gt is present in a narrower anterior domain and Kr is present in a central stripe. Gt antagonizes activation by Hb and Bcd to define the anterior border of stripe 2, while Kr antagonizes activation by Hb and Bcd to define the posterior border of stripe 2.

and two of which direct the formation of two stripes each (Figure 7.5A). The best understood CRM in *eve* is the one that directs formation of the second *eve* stripe (Figure 7.5B). This CRM contains multiple binding sites for the activators Hunchback and Bicoid, which are both distributed in a broad domain roughly corresponding to the anterior half of the embryo. The Hunchback and Bicoid sites are interspersed with sites for the two repressors Giant and Kruppel. In the absence of the repressors, Hunchback and Bicoid, which work synergistically with one another, would be expected to bind the stripe 2 enhancer and activate *eve* throughout the anterior half of the embryo. However, the repressors serve to delimit this domain of activation. Giant, which is found in an anterior domain, inhibits Hunchback and Bicoid activity to set the anterior border of stripe 2, while Kruppel, with is found in a central domain, inhibits Hunchback and Bicoid activity to set the posterior border of stripe 2 (Figure 7.5C).

Thus, the definition of this single stripe of transcription requires at least two spatially restricted activators and two spatially restricted repressors, not to mention a host of coactivators and corepressors that will not be considered here. Similarly elaborate sets of interactions between activators, repressors, and coregulators acting through four additional stripe enhancers are required to define each of the other six stripes of *eve* expression. This seemingly inelegant system for generating stripes is reminiscent of a Rube–Goldberg device, which is a device for achieving minimal results with maximal effort (see http://www.rube-goldberg.com for a gallery of Rube–Goldberg devices). Like a Rube–Goldberg device, we might expect the stripe-forming system to be extremely sensitive to any changes in the environment (e.g., changes in temperature or humidity) that could alter the activity of one or more of the factors required for stripe formation. Contrary to this expectation, the system is resistant to perturbation – the seven equally spaced and equally intense stripes of expression form with amazing reliability. We do not understand the basis for this reliability, but it suggests that the stripe-forming system must be able to compensate for environmental perturbation. It may contain self-organizing features that allow it to adjust the activity of one factor to compensate for changes in the activity of another.

Enhancer autonomy

Repressors such as Giant and Kruppel that regulate stripe formation are short-range repressors. By mechanisms that we do not understand, they interfere with the function of activators bound nearby (within ~100 bp of the repressors), but not with the function of more distantly bound activators. Short-range repression prevents stripe enhancers from interfering with one another's activity, i.e., short-range repression allows the stripe enhancers to work as autonomous units.

The importance of enhancer autonomy in the overall regulation of *eve* expression is best illustrated by experiments examining the *eve* stripe 2 and stripe 3 + 7 enhancers. These two enhancers are normally separated by a thousand basepairs. This distance is too long to allow short-range repressors bound to the stripe 2 enhancer to antagonize activators bound to the stripe 3 + 7 enhancer. This is essential because Kruppel, which defines the posterior border of stripe 2, is distributed in a relatively broad domain that includes the region in which stripe 3 forms. If Kruppel bound to the stripe 2 enhancer was able to

> **What determines the distance of repressor action?**
>
> While antagonism between activators (or coactivators) and repressors (or corepressors) is absolutely essential in combinatorial control, we have a very poor understanding of what determines whether or not one factor will be able to antagonize another. For example, we do not understand why some regulatory factors (e.g., PRC1) are able to repress over long distances, while others (e.g., Kruppel) are only able to repress over short distances.

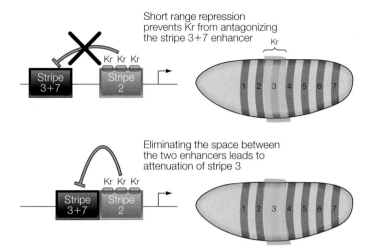

Figure 7.6 *Enhancer autonomy*. The *eve* stripe 2 and stripe 3 + 7 enhancers are normally separated by about 1000 bp. Since Kruppel (Kr) is a short range repressor, Kr bound to the stripe 2 enhancer is unable to antagonize activators bound to the stripe 3 + 7 enhancer. This is important because stripe 3 is wholly contained within the central Kr stripe. If the spacer between the two enhancers is eliminated, Kr is then able to antagonize activators bound to the stripe 3 + 7 enhancer leading to a significant weakening of stripe 3.

block activation by activators bound to the stripe 3 + 7 enhancer, the result would be the inappropriate weakening of stripe 3 (Figure 7.6).

In summary, the expression of pair-rule genes in periodic stripes requires multiple CRMs, each of which carries the instruction for one or two stripes. Formation of these stripes requires activators, which turn genes on in broad domains, as well as repressors, which antagonize the activators to set the borders of the stripes. The antagonistic effects of the repressors are limited to activators bound within the same CRM, allowing each stripe enhancer to function autonomously. The mechanism of this short-range repression remains to be determined. One likely possibility is that each stripe enhancer mediates formation of an enhanceosome and that the repressors act in a local manner to disrupt the formation or function of these enhanceosomes, perhaps by interfering with coactivator recruitment.

7.5 ANTAGONISM AND THE SIGNAL-MEDIATED SWITCH

A highly prevalent form of combinatorial control in signal transduction is the signal-mediated switch, in which an extracellular signal serves to switch a gene on. Many of these switches can be assigned to one of

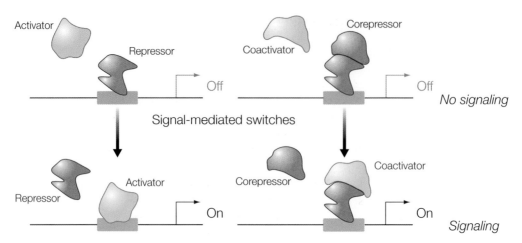

Figure 7.7 *Antagonism and signal-mediated switches.* Environmental signals are often relayed to the nucleus where they switch genes from an off state to an on state. This often involves antagonism between activators and repressors competing for the same DNA recognition element, or between coactivators and corepressors competing for the same protein interaction surface. In the absence of the signal, the repressor or corepressor has the advantage in the competition, while, in the presence of the signal, the activator or coactivator gains the advantage.

two categories (Figure 7.7). First, the switches sometimes involve competition between activators and repressors for the same cis-regulatory elements in the DNA. Second, they sometimes involve competition between coactivators and corepressors for protein interaction surfaces on sequence-specific factors. In each case, the repressor or corepressor has the advantage in the competition in the absence of the signal and thus the target genes are off. In the presence of the signal, however, the activator or coactivator gains the upper hand in the competition turning the target genes on. The remainder of this chapter includes two examples of the signal-mediated switch with widespread importance in multicellular eukaryotes. The first example, the nuclear receptor-mediated switch, represents a case in which coactivators and corepressors compete for a common protein interaction surface, while the second example, the receptor tyrosine kinase-mediated switch, represents a case in which activators and repressors compete for the same DNA recognition element.

Since eukaryotic core promoters are generally inactive in the absence of activators (see Chapter 4), it might seem unnecessary to employ a repressor to keep a gene off in the absence of a signal. However, as will be made clear, genes that are activated by signal transduction pathways are frequently targets of repression in the absence of the signal. This apparent redundancy is essential to ensure the strict control required for stringent responses to signals.

7.5.1 Nuclear receptors: antagonism between coregulators

Overview of nuclear receptors

The picture that most often comes to mind when one thinks of a receptor is that of a transmembrane protein containing an extracellular ligand-binding domain and an intracellular effector domain. However, nuclear receptors represent important exceptions to this rule. These intracellular receptors bind directly to lipid-soluble ligands, such as steroid hormones, fatty acids, thyroid hormone, retinoids (vitamin A derivatives), and vitamin D, which move freely across cell membranes. In response to these ligands, they then function as transcriptional activators, turning on a variety of target genes often leading to profound changes in cellular physiology.

The nuclear receptors with known ligands can be divided into two classes: class I includes all the steroid hormone receptors, while class II includes the receptors known to bind non-steroidal ligands (Figure 7.8). In addition to the receptors with known ligands, a large number of nuclear receptor family transcription factors do not have known ligands and are therefore often described as "orphan receptors". In general, the properties of orphan receptors are more similar to those of the class II receptors than to those of the class I receptors.

All nuclear receptors bear a conserved zinc-containing DNA-binding domain (see Chapter 4). In addition, nuclear receptors usually bear a conserved C-terminal ligand-binding domain (Figure 7.8A). Most nuclear receptors function as dimers, with dimerization being mediated by interfaces in both the DNA-binding and ligand-binding domains. Class I receptors form symmetric homodimers that bind to recognition elements consisting of two half-sites arranged in inverted repeat orientation (Figure 7.8B). Class II receptors (and many orphan receptors) heterodimerize with a common partner, the retinoid X receptor (RXR). These heterodimers are stabilized by a head to tail interaction between the DNA-binding domains of the two partners and bind to directly repeated half-sites (Figure 7.8C).

The DNA recognition helix in the class II nuclear receptors is highly conserved. As a result, most class II receptors recognize the same half-site. However, as a consequence of subtle variations in the heterodimeric interfaces, different receptors prefer half-sites spaced at different distances (Figure 7.8C). This represents a form of combinatorial diversity, as it provides a way of using a conserved recognition helix to bind different recognition elements in response to different signals.

Allosteric regulation of ligand-binding domain conformation

Nuclear receptor ligand-binding domains provide a classic example of allosteric regulation, in which a ligand (an "allosteric effector") binds to

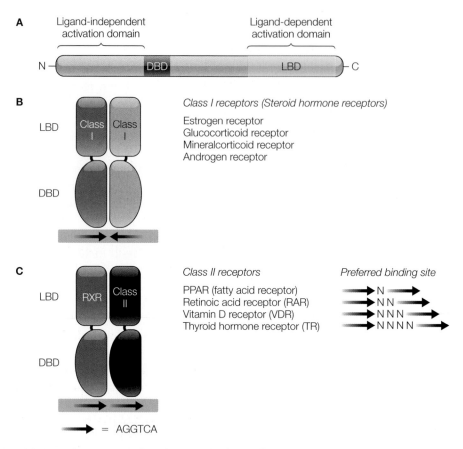

Figure 7.8 *Nuclear receptor domain organization and DNA recognition.* (A) Nuclear receptors contain a central DNA-binding domain (DBD) and a C-terminal ligand-binding domain (LBD). The LBD functions as a ligand-dependent activation domain. In addition, the poorly conserved region N-terminal to the DBD often functions as a ligand-independent activation domain. (B) Class I nuclear receptors form symmetric homodimers and bind to inverted repeats. Class II nuclear receptors usually heterodimerize with the retinoid X receptor (RXR). The asymmetric heterodimers bind to directly repeated half-sites. Half-site spacing varies depending on the identity of the class II receptor. The RXR:PPAR heterodimer prefers a 1 bp spacer, the RXR:RAR heterodimer prefers a 2 bp spacer, the RXR:VDR heterodimer prefers a 3 bp spacer, and the RXR:TR heterodimer prefers a 4 bp spacer.

one conformation of a protein but not another, thereby favoring the conformation to which it binds. Structural studies show that the overall fold of the ligand-binding domain is highly conserved among nuclear receptors. This fold is sometimes described as a helical sandwich because it contains 12 α-helices arranged in three layers (Plate 7.1).

Binding of the ligand to the ligand-binding domain results in an important change in the conformation of the domain (Plate 7.1A). In the ligand-bound receptor, the C-terminal most α-helix (helix 12) forms one wall of the ligand-binding pocket. In the unbound receptor, however, this helix is conformationally mobile, frequently moving away from the rest of the domain and therefore leaving the binding pocket open. Once the ligand binds in the pocket, favorable interactions between the ligand and helix 12 firmly cement helix 12 in place over the binding pocket, sealing the ligand off from the solvent. In effect, helix 12 functions like a trap door, which the ligand pulls shut after it enters the binding pocket.

Interactions between nuclear receptors and coactivators

The above-described conformational change leads to the activation of nuclear receptor targets by increasing the affinity of nuclear receptors for coactivators. A number of coactivators that bind nuclear receptors, including the Mediator complex (see Chapter 4), contain multiple copies of a leucine-rich motif with the sequence LXXLL. High resolution structural studies reveal that the LXXLL motif forms a short α-helix that binds in a groove on the ligand-binding domain surface. The N-terminal end of this helix (the positive end of the α-helix dipole) is held in position by H-bond contacts with a conserved, negatively charged glutamate side chain in helix 12 of the ligand-binding domain, while the C-terminal end (the negative end of the α-helix dipole) is held in position by H-bond contacts with a conserved, positively charged lysine side chain protruding from another α-helix (Plate 7.1A). Only when helix 12 is positioned over the ligand-binding pocket (i.e., only when the trap door is shut) is the glutamate side chain positioned to contact the N-terminus of the LXXLL

> **α-helix dipole** – the asymmetric charge distribution in α-helices that results from having all the peptide bonds pointing in the same direction. The N-terminal end of the helix has a positive charge, while the C-terminal end has a negative charge.

helix. Thus, the repositioning of helix 12 that accompanies ligand binding results in the formation of a "charge clamp" capable of gripping the LXXLL motif.

The biological significance of helix 12 movement is illustrated by studies of the anticancer drug 4-hydroxytamoxifen (OHT), an estrogen receptor antagonist used as a highly effective therapy for breast cancer. OHT binds in the ligand-binding pocket of the estrogen receptor, assuming a position similar to that assumed by estrogen (Plate 7.1B). However, OHT is too bulky to allow helix 12 to assume its normal position (that is, OHT blocks closure of the trap door). Instead, helix 12 binds in the groove normally occupied by the LXXLL coactivator motif. As a result, coactivators are excluded from the groove preventing estrogen-dependent activation of the genes that drive the proliferation of breast cancer cells.

Competition for a common surface by coactivators and corepressors

In addition to the ligand-dependent coactivator-binding surface in the ligand-binding domain, most nuclear receptors contain additional activation domains that are *not* directly regulated by ligand binding (see Figure 7.8A). These ligand-independent activation domains could theoretically lead to aberrant ligand-independent transcription of nuclear receptor target genes. Making this potential problem even worse, nuclear receptors generally bind CRMs that contain binding sites for many additional activator proteins that are not ligand regulated. The combined effects of ligand-independent activation domains in nuclear receptors and ligand-independent activation domains in other activators might be expected to result in high levels of nuclear receptor target gene expression even in the absence of the relevant ligand.

Class II nuclear receptors have evolved a strategy to block this ligand-independent transcription of their target genes. In the absence of ligand, they recruit a corepressor termed NCoR, which contains two copies of a short nuclear receptor interaction motif with the consensus sequence LXXXIXXXI/L. This motif is reminiscent of the LXXLL motif that mediates the binding of coactivators to nuclear receptor ligand-binding domains, and it is believed that the corepressor and coactivator motifs bind to the same groove on the surface of the ligand-binding domain in a mutually exclusive manner. However, unlike the coactivators, which bind preferentially to liganded receptors, the corepressors bind preferentially to unliganded receptors. This difference in preference probably results from the greater length of the LXXXIXXXI/L motif relative to the LXXLL motif. As a result of this length, the corepressor motif cannot fit between the critical charged residues defining the charge clamp and is thus expected to bind poorly to the liganded receptor.

Once recruited to the unliganded receptor, NCoR broadly interferes with transcriptional activation. It probably does so by recruiting the Sin3 histone deacetylase complex (see Chapter 5) to nuclear receptor target genes where it catalyzes deacetylation of histone tails and thus the organization of chromatin into a transcriptionally silent conformation.

In conclusion, conformational changes brought about by ligand binding influence the ability of coactivators and corepressors to compete for the same interaction surface on nuclear receptor ligand-binding domains. In the absence of ligand, nuclear receptors recruit corepressors leading to repression. This ensures that target gene activity is kept extremely low in the absence of signaling.

7.5.2 Receptor tyrosine kinase pathways: competition for a common DNA element

Rather than using a single sequence-specific factor to mediate both activation and repression, the receptor tyrosine kinase (RTK) pathway employs multiple sequence-specific factors that compete for the same binding sites in the target genes. RTKs are transmembrane proteins containing an extracellular ligand-binding domain and an intracellular protein tyrosine kinase domain. RTKs mediate responses to protein ligands, particularly growth factors, including epidermal growth factor, nerve growth factor, and fibroblast growth factor. Binding of one of these ligands to its cognate RTK stimulates an intracellular protein kinase cascade leading to the phosphorylation and activation of mitogen activated protein (MAP) kinase, which, in turn, phosphorylates multiple transcription factors altering their properties (Figure 7.9).

Yan, a repressor, and Pnt-P2, an activator, are two of the many factors that are phosphorylated by MAP kinase. Both of these factors contain similar DNA-binding domains of the helix-turn-helix class (see Chapter 4). As a result of the homology between the Yan and Pnt-P2 DNA-binding domains, they compete for the same cis-regulatory elements in RTK pathway target genes.

In the absence of signals that stimulate the activity of the RTK pathway, Yan and Pnt-P2 remain unphosphorylated. In the absence of phosphorylation, Yan (the repressor) is favored over Pnt-P2 (the activator) in the competition for cis-regulatory elements in the RTK pathway target genes. Thus these genes are maintained in the off state. Upon stimulation of the RTK pathway, phosphorylation of Yan and Pnt-P2 by MAP kinase tips the balance in favor of Pnt-P2 and consequently the target genes transition from the off state to the on state.

How does phosphorylation alter the competition in favor of Pnt-P2? In general, phosphorylation of transcription factors can influence their activity in many ways, for example, by altering their stability, their subcellular localization, or their ability to bind coregulators. Phosphorylation of Yan results in the recognition of Yan by the nuclear export machinery, which then escorts Yan out of the nucleus into the cytoplasm where it is unable to repress transcription. In addition, phosphorylation of Pnt-P2 may increase its affinity for the coactivator CBP (see above section on enhanceosomes), leading to efficient recruitment of this coactivator and therefore efficient activation of target genes.

7.6 SUMMARY

Complex organisms employ combinatorial control to achieve far more diverse patterns of gene expression than would otherwise be possible. In this type of control, multiple sequence-specific transcription factors and coregulators interact to determine the transcriptional state of a gene. Two major forms of combinatorial control used by all organisms are synergy, in which multiple activators collaborate to direct high levels of transcription, and antagonism, in which repressors block the function of activators. Synergy provides a way to ensure that a gene will only be

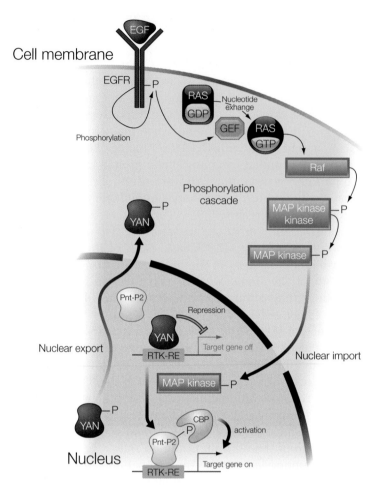

Figure 7.9 *Antagonistic regulation of receptor tyrosine kinase target genes.*
Epidermal growth factor (EGF) signals through a receptor tyrosine kinase,
the EGF receptor (EGFR). EGF holds two EGFR monomers together
allowing the cytoplasmic domain of one to phosphorylate tyrosine residues
in the cytoplasmic domain of the other. The phosphorylated cytoplasmic
domain then stimulates guanine nucleotide exchange factor (GEF). This
protein stimulates release of GDP from the Ras protein, which can then
bind GTP. The GTP-bound form of Ras stimulates a protein kinase termed
Raf, and the initiation of a phosphorylation cascade involving the
serine/threonine protein kinases Raf, MAP (mitogen associated protein)
kinase kinase, and MAP kinase. Phosphorylated MAP kinase enters the
nucleus and phosphorylates a number of transcription factors including
Yan and Pnt-P2. These two factors compete for the same RTK (receptor
tyrosine kinase) response elements (RTK-REs) in various RTK target genes.
In the absence of phosphorylation, Yan is favored in the competition and
serves to repress gene expression. Upon phosphorylation of Yan by MAP
kinase, Yan is recognized by the nuclear export machinery and therefore
exported from the nucleus. This allows Pnt-P2 to bind the RTK-REs and
turn on gene expression. Phosphorylation of Pnt-P2 appears to facilitate
recruitment of the coactivator CBP.

active in the presence of multiple inputs, while antagonism provides a means of weighing both positive and negative inputs in making a regulatory decision.

Enhanceosomes, which provide a mechanistic explanation for synergy, are cooperatively assembling nucleoprotein complexes that form at enhancers. These complexes contain multiple activators and coactivators, and their assembly depends on architectural factors that alter the curvature of DNA, thereby facilitating cooperative interactions between DNA-bound factors. Activators bound to adjacent sites in enhanceosomes are thought to form high affinity platforms for the recruitment of coactivators. Enhanceosomes often recruit multiple coactivators in a sequential manner leading to transcriptional activation.

Antagonism plays a central role in the spatial regulation of gene expression during development. This is exemplified by the mechanisms that direct the expression of *Drosophila* pair-rule genes such as *eve* in seven regularly repeating stripes along the length of the embryo. The formation of these stripes requires the action of complex sets of CRMs; *eve*, for example, contains five CRMs (termed stripe enhancers), each of which directs one or two of the seven stripes of *eve* expression. These stripe enhancers each interact with multiple activators and repressors. The activators turn the promoter on in broad domains, while the repressors antagonize the activators to set the borders of the stripes. The repressors act by short-range mechanisms, thus ensuring that each stripe enhancer will function as an autonomous unit.

Antagonism is also essential for signal-mediated transcriptional switches. These switches often occur when a signal influences the competition between an activator and a repressor (or a coactivator and a corepressor) in favor of the activator (or coactivator).

Nuclear receptors represent one type of signal-mediated switch. These transcription factors contain surfaces in their ligand-binding domains that interact with both coactivators and corepressors. Binding of ligand induces a conformational change that favors the coactivator interaction leading to target gene activation. In the absence of the ligand, corepressors bind to the ligand-binding domain and enforce repression.

The receptor tyrosine kinase pathway represents another type of signal-mediated switch. In this pathway, activator and repressor proteins compete for common DNA recognition elements. In the presence of a signal, these receptors trigger an intracellular cascade of protein phosphorylation reactions leading to the phosphorylation of multiple transcription factors, including the repressor Yan and the activator Pnt-P2. Prior to phosphorylation, Yan prevails in the competition with Pnt-P2 for DNA recognition elements and therefore target genes remain off. Upon phosphorylation, Yan is exported from the nucleus and thus Pnt-P2 is able to bind the recognition elements and activate the target genes.

PROBLEMS

1 How would elimination of Kruppel (Kr) from the embryo change the domain of expression directed by the *eve* stripe 2 enhancer?

2 As shown in Figure 7.6, removing the spacer between the stripe 2 enhancer and the stripe 3 + 7 enhancer weakens stripe 3, but not stripe 7. Why is stripe 7 not weakened?

3 OHT and other related antagonists of estrogen receptors (ERs) are sometimes termed partial antagonists. This is because, although they block estrogen function, they do not completely inactivate ER target genes. Based on the structure of the OHT–ER complex (Plate 7.1B), why do you think OHT fails to inactivate ER target genes completely?

FURTHER READING

Enhanceosomes

Thanos, D. and Maniatis, T. (1995) Virus induction of human IFN beta gene expression requires the assembly of an enhanceosome. *Cell*, **83**, 1091–1100. *Demonstrates that alignment of transcription factors on the same face of the helix is required for enhanceosome function; shows that HMG I has an architectural function.*

Merika, M., Williams, A.J., Chen, G., Collins, T. and Thanos, D. (1998) Recruitment of CBP/p300 by the IFN beta enhanceosome is required for synergistic activation of transcription. *Mol Cell*, **1**, 277–287. Agalioti, T., Lomvardas, S., Parekh, B., Yie, J., Maniatis, T. and Thanos, D. (2000) Ordered recruitment of chromatin modifying and general transcription factors to the IFN-beta promoter. *Cell*, **103**, 667–678. *Two papers demonstrating cooperative coactivator recruitment and showing that enhanceosomes recruit multiple coactivators in sequence.*

Formation and interpretation of the Bicoid transcription factor gradient

Driever, W. and Nusslein-Volhard, C. (1988) A gradient of bicoid protein in Drosophila embryos. *Cell*, **54**, 83–93. *Establishment of a transcription factor gradient by diffusion from a source.*

Driever, W., Thoma, G. and Nusslein-Volhard, C. (1989) Determination of spatial domains of zygotic gene expression in the Drosophila embryo by the affinity of binding sites for the bicoid morphogen. *Nature*, **340**, 363–367. *Interpretation of a transcription factor gradient with the use of different affinity binding sites.*

Stripe formation

Akam, M. (1989) Drosophila development: making stripes inelegantly. *Nature*, **341**, 282–283. *A review article lamenting the inelegant mechanism of stripe formation.*

Small, S., Blair, A. and Levine, M. (1992) Regulation of even-skipped stripe 2 in the Drosophila embryo. *EMBO J*, **11**, 4047–4057. Gray, S. and Levine, M. (1996) Short-range transcriptional repressors mediate both quenching and direct repression within complex loci in Drosophila. *Genes Dev*, **10**, 700–710. *Two papers showing how short-range repression is used in stripe formation.*

Small, S., Arnosti, D.N. and Levine, M. (1993) Spacing ensures autonomous expression of different stripe enhancers in the even-skipped promoter. *Development*, **119**, 762–772. *The importance of short-range repression in ensuring enhancer autonomy.*

Clyde, D.E., Corado, M.S., Wu, X., Pare, A., Papatsenko, D. and Small, S. (2003) A self-organizing system of repressor gradients establishes segmental complexity in Drosophila. *Nature*, **426**, 849–853. *Demonstration of the elegant self-organizing nature of the eve stripe-forming system.*

Nuclear receptors

Umesono, K., Murakami, K.K., Thompson, C.C. and Evans, R.M. (1991) Direct repeats as selective response elements for the thyroid hormone, retinoic acid, and vitamin D3 receptors. *Cell*, **65**, 1255–1266. Perlmann, T., Rangarajan, P.N., Umesono, K. and Evans, R.M. (1993) Determinants for selective RAR and TR recognition of direct repeat HREs. *Genes Dev*, **7**, 1411–1422. *Two papers showing that half-site spacing is the determinant of class II nuclear receptor specificity.*

Shiau, A.K., Barstad, D., Loria, P.M., Cheng, L., Kushner, P.J., Agard, D.A. and Greene, G.L. (1998) The structural basis of estrogen receptor/coactivator recognition and the antagonism of this interaction by tamoxifen. *Cell*, **95**, 927–937. Nolte, R.T., Wisely, G.B., Westin, S., Cobb, J.E., Lambert, M.H., Kurokawa, R., Rosenfeld, M.G., Willson, T.M., Glass, C.K. and Milburn, M.V. (1998) Ligand binding and co-activator assembly of the peroxisome proliferator-activated receptor-gamma. *Nature*, **395**, 137–143. *Two papers demonstrating the allosteric regulation of nuclear receptor ligand-binding domain function.*

Heinzel, T., Lavinsky, R.M., Mullen, T.M., Soderstrom, M., Laherty, C.D., Torchia, J., Yang, W.M., Brard, G., Ngo, S.D., Davie, J.R., Seto, E., Eisenman, R.N., Rose, D.W., Glass, C.K. and Rosenfeld, M.G. (1997) A complex containing N-CoR, mSin3 and histone deacetylase mediates transcriptional repression. *Nature*, **387**, 43–48. Nagy, L., Kao, H.Y., Chakravarti, D., Lin, R.J., Hassig, C.A., Ayer, D.E., Schreiber, S.L. and Evans, R.M. (1997) Nuclear receptor repression mediated by a complex containing SMRT, mSin3A, and histone deacetylase. *Cell*, **89**, 373–380. *Two papers establishing the idea that nuclear receptor corepressors function to recruit the Sin3 HDAC complex.*

Receptor tyrosine kinase signaling

O'Neill, E.M., Rebay, I., Tjian, R. and Rubin, G.M. (1994) The activities of two Ets-related transcription factors required for Drosophila eye development are modulated by the Ras/MAPK pathway. *Cell*, **78**, 137–147. Gabay, L., Scholz, H., Golembo, M., Klaes, A., Shilo, B.Z. and Klambt, C. (1996) EGF receptor signaling induces pointed P1 transcription and inactivates Yan protein in the Drosophila embryonic ventral ectoderm. *Development*, **122**, 3355–3362. *Two papers demonstrating antagonism between the two ets family factors Yan and Pnt-P2.*

Tootle, T.L., Lee, P.S. and Rebay, I. (2003) CRM1-mediated nuclear export and regulated activity of the receptor tyrosine kinase antagonist YAN require specific interactions with MAE. *Development*, **130**, 845–857. *Proof that phosphorylation inactivates Yan by leading to its export from the nucleus.*

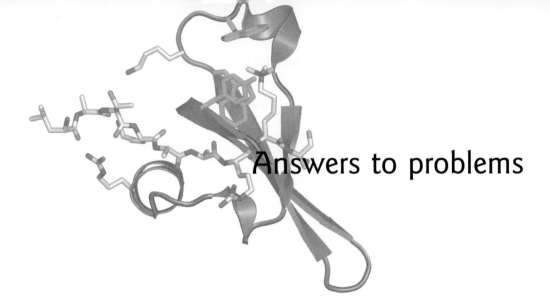

Answers to problems

CHAPTER 1

1 As discussed in the paper by Levine and Tjian (2003; see further reading in Chapter 1), the main driver of evolutionary diversity may be the evolution of transcriptional regulation. If the same machinery directed catalysis and regulation of transcription, this might slow down the evolution of regulatory diversity since mutations that alter regulation might inactivate the catalytic machinery. Thus, the existence of separate machinery for regulation may speed up this evolution allowing organisms to adapt more readily to a changing environment.

2 The review by Ptashne (1988; see further reading in Chapter 1) suggests a number of possibilities.

CHAPTER 2

1 Like phosphoester bond formation, hydrolysis requires two metal ions. But in the absence of nucleotides, only one of the two metal ions (metal A) is firmly bound in the active site. An exception to this occurs when an elongation factor such as GreA, GreB, or TFIIS binds in the secondary channel. These factors use conserved, negatively charged amino acid side chains to position the second metal ion (metal B) in the active site.

2 By destabilizing the heteroduplex, errors could lead to backtracking. Resolution of the backtracked complex by hydrolysis would then serve to remove the erroneous nucleotides.

3 The presence of microcin J25 in the secondary channel could interfere with the binding of factors such as GreA or GreB and thereby prevent resolution of the backtracked complex.

4 Early during transcription, when the heteroduplex is short, fidelity requires tight packing of the incoming nucleotide in the active site to prevent the formation of illegitimate basepairs (non-Watson–Crick basepairs) between the nucleotide and the template. However, as the

transcript grows in length, the rigidity of the long heteroduplex is sufficient to ensure proper positioning of the nucleotide in the active site. At this stage, the tight packing between the enzyme and the substrates might serve to prevent the 5′ end of the nascent transcript from escaping the active site. Bacterial RNA polymerase has resolved this conflict by placing a segment of the σ factor (the σ_3–σ_4 linker) in the active site to provide a part of the nucleotide-binding interface during the early elongation cycles. Then as the elongation proceeds, the 5′ end of the transcript pushes the σ_3–σ_4 linker out of the way to provide an exit channel for the transcript.

5 The holoenzyme structure shows that domain 4 of the σ subunit binds to the tip of the flap. In the absence of the flap, domain 4 of σ will not be properly positioned with respect to the core, thus interfering with closed complex formation.

6 The low stability of I:C basepairs will reduce the stability of the hairpin that normally forms in the transcript after transcription through the termination signal. This hairpin is what normally triggers termination.

CHAPTER 3

1 The need for a separate enzyme to catalyze open complex formation provides an additional target for regulation.

2 (a) One possible approach would be to mutate critical residues in TBP required for DNA binding and determine if these mutant forms of TBP are still able to support transcription of TATA-less promoters in a reconstituted *in vitro* system. (b) In the absence of a TATA box, TFIID probably binds to the promoter via interactions between TAFs and other core promoter elements such as the inr or the DPE. This brings TBP to the promoter so that it can recruit TFIIB, which acts as the bridge to Pol II.

3 Both TFIIF and σ bind to core polymerase in the absence of DNA.

CHAPTER 4

1 The specificity of protein–DNA interactions is largely provided by specific H-bond interactions. In other words, specific binding occurs when a surface on a protein (such as one face of a recognition helix) contains an array of H-bond acceptors and donors that is complementary to an array of H-bond donors and acceptors found in one of the grooves of a segment of DNA. The asymmetric pattern of H-bond donors and acceptors in the major groove means that proteins binding in the major groove can distinguish GC from CG basepairs and AT from TA basepairs. In contrast, the minor groove displays a symmetric array of

H-bond donors and acceptors and, as a result, proteins binding in the minor groove cannot distinguish GC from CG or AT from TA basepairs. These ideas are discussed in the paper by Seeman et al. (1976; see further reading in Chapter 4).

2 Since the AR1 mutation lowers K_B, the role of AR1 is to stimulate recruitment of RNA polymerase to the closed complex. Since the AR2 mutation lowers k_f, the role of AR2 is to stimulate conversion of the closed complex to the open complex. Note that the role of AR2 in open complex formation does not conflict with the data in Table 4.1, which shows that CRP is not required for open complex formation at the lac promoter. This is because AR2 plays no role in the activation of the lac promoter.

3 The genes that gave rise to the dominant suppressors all encode subunits of the head module, while the genes that gave rise to the recessive suppressors all encode subunits of the Cdk8 module. The dominant mutations of SRB2, SRB4, SRB5, and SRB6 could work by strengthening the interaction between the head module and Pol II, thus allowing the Mediator to function even in the presence of a severely truncated CTD. The recessive mutations in SRB8–SRB11 might inactivate the Cdk8 module. Recall that this module is thought to inhibit Mediator function. Inactivation of this module might therefore counteract the effect of the CTD truncation.

CHAPTER 5

1 ATP-dependent chromatin remodelers are thought to induce the dissociation of DNA from a part of the histone octamer. By destabilizing the octamer, this could lead to further nucleosome disassembly.

2 Both of these lysine residues are embedded in the same sequence context within the histone H3 N-terminal tail (Ala-Arg-Lys-Ser), suggesting that it would be difficult for chromodomains to distinguish methylated Lys9 from methylated Lys27. However, inspection of the crystal structure reveals contacts between the chromodomain and the histone H3 tail outside of this conserved motif. Specifically, the crystal structure of the complex between the H3 tail and the Polycomb chromodomain shows an H-bond between Lys48 of the chromodomain and the threonine side chain five amino acids before Lys27 of histone H3. Note that Arg67 of the Polycomb chromodomain makes contacts with backbone carbonyl oxygens in the vicinity of this threonine residue that should help to position the threonine residue for the sequence-specific interaction with Lys48. The residue five amino acids before Lys9 is a lysine and not a threonine. It would therefore not make the same contact.

3 The ChIP on chip results suggest that histone H4 Lys16 acetylation generally leads to transcriptional repression. On the other hand, the

role of SAS in blocking heterochromatin formation would seem to suggest a role for SAS in activation. There is no easy way to reconcile these discordant observations.

4 You might expect to recover many other components of the regulatory machinery required for activation by the VP16 activation domain, including Mediator subunits. The failure to find these other regulatory factors suggests that they are essential for viability, perhaps due to their broad roles in transcription.

CHAPTER 6

1 In the model presented in this chapter, PRC1 (the complex containing Pc) is envisioned to inhibit ATP-dependent remodeling by Brahma. Thus, in the absence of Brahma, Pc lacks a regulatory target. Therefore we expect the double mutant to have the same phenotype as the *brahma* single mutant (failure to maintain the active state of the homeotic genes).

2 The recruitment of Ash1 by the TRE transcripts might set up a self-maintaining feedback cycle. When Ubx, and therefore the associated TRE, is transcriptionally active, this would lead to recruitment of Ash1 by the nascent transcripts, which would methylate histone H3 lysine 4, stabilizing the active state.

3 Both mutations would reduce levels of histone H3 lysine 9 di- and trimethylation in centromeric chromatin. The CLR4 mutation would also reduce methylation in the *K*-region. However, the Argonaut mutation might have little or no effect on methylation in the *K*-region. This is because there are dsRNA-independent pathways involving sequence-specific transcription factors for recruitment of CLR4 to the *K*-region.

4 The calico cat contains large patches of orange and black fur. This suggests that once an X chromosome is selected for inactivation during early embryogenesis, this state is maintained throughout the remaining cell divisions giving rise to the mature animal.

CHAPTER 7

1 We would expect to see expansion of stripe 2 toward the posterior of the embryo.

2 Gt and Kr, the repressors that block activation by the stripe 2 enhancer to set the anterior and posterior borders of stripe 2, are not present in the part of the embryo where stripe 7 forms. So removing the spacer has no effect on stripe 7. The mechanism by which the stripe 3 + 7 enhancer directs the formation of two stripes is presented in the paper by Clyde et al. (2003; see further reading in Chapter 7).

3 When OHT binds to the ligand-binding domain, helix 12 occupies the position normally occupied by the LXXLL motif found in many coactivators. Corepressors such as NCoR are thought to bind to this same surface using an LXXXIXXXI/l motif. Thus, OHT blocks the recruitment of both coactivators and corepressors. As a result, ligand-independent activation domains in estrogen receptors as well as ligand-independent activation domains in other activators that bind to nearby sites in the DNA are still able to activate transcription in the presence of OHT.

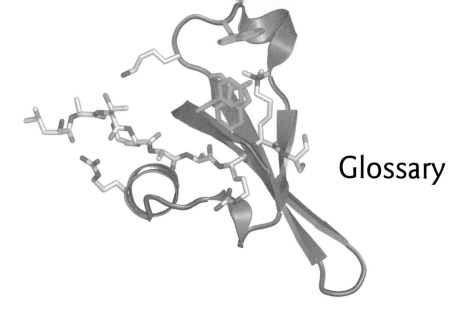

Glossary

abortive initiation assay – a method for monitoring the rate of open complex formation and thereby measuring the energetic and kinetic parameters such as K_B and k_f that govern open complex formation (Box 4.3).

acidic activation domain – *see* activation domain.

activation domain – protein domains found in eukaryotic activators that signal either directly or indirectly to the basal transcriptional machinery to stimulate transcription. They bind to various targets in the transcriptional machinery including general transcription factors, the Mediator, and histone acetyltransferases. Activation domains appear to be poorly ordered when not bound to their targets. They are sometimes characterized by an abundance of glutamate and aspartate residues in which case they are termed acidic activation domains (p. 89).

activator – *see* sequence-specific transcription factor.

adenylate cyclase – *see* cyclic AMP (cAMP).

allosteric regulation – a form of regulation that depends on the ability of protein ligands to stabilize one protein conformation in favor of another. If a ligand binds favorably to one conformation of a protein and not another, the favorable free energy change that results from ligand binding will stabilize the conformation of the protein to which the ligand binds relative to the form of the protein to which the ligand is unable to bind. If the ligand binds exclusively to the active form of the protein, then the ligand increases protein activity by stabilizing the active form of the protein relative to the inactive form of the protein. Examples of allosteric regulation discussed in this book include activation of RNA polymerase by λcI (p. 85) and activation of nuclear receptors by their ligands (p. 185).

α-amanitin – a cyclic octapeptide produced naturally by the mushroom *Amanita phalloides* (the death cap mushroom) that binds and inhibits eukaryotic RNA polymerases. Pol II is exquisitely sensitive to this enzyme, Pol I is insensitive, and Pol III exhibits intermediate sensitivity (p. 19).

α-helix dipole – the dipole associated with every α-helix that results from all the peptide bonds pointing in the same direction. The positive end of the dipole is at the N-terminal end of the helix, while the negative end of the dipole is at the C-terminal end of the helix (p. 187).

amphipathic α-helix – an α-helix in which one face is dominated by hydrophobic (water-hating) non-polar amino acid side chains and the other face by hydrophilic (water-loving) polar amino acid side chains. The hydrophobic faces of two such helices often bind to one another in such a way as to shield these hydrophobic side chains from the watery environment. Examples of such interactions are provided by the leucine zippers found in bZip domains (p. 87).

AND operator – *see* synergy.

antagonism – a form of combinatorial control in which a repressor (or corepressor) blocks the function of activators (or coactivators). In mathematical terms, antagonism can be described with the use of the NOT operator since gene activity requires the presence of activator X, but the absence of repressor Y, i.e., fulfillment of the condition X NOT Y (p. 169).

Antennapedia (Antp) – a *Drosophila* homeodomain-containing transcription factor encoded by one of the genes of the homeotic gene complex (p. 151).

AP-1 – a human bZip domain-containing transcription factor related to the yeast factor Gcn4 (p. 87).

architectural factor – *see* enhanceosome.

Argonaut – *see* RNA interference (RNAi).

Ash1 – a member of the Trithorax group and a histone lysine methyl-transferase that trimethylates histone H3 lysine 4 (p. 157).

Atf1 – a bZip domain-containing transcription factor. Along with Pcr1 (another bZip domain-containing factor), it binds to the silencers in the fission yeast silent mating type loci and initiates heterochromatin formation by recruiting components of the heterochromatin forming machinery such as Clr4 and Swi6 (p. 146).

ATF-2/c-Jun – a bZip domain-containing heterodimeric transcription factor that is a component of the IFNβ enhanceosome (p. 173).

ATP-dependent chromatin remodeling complex – *see* chromatin remodeling.

backtracking – the process in which RNA polymerase engaged in elongation stalls and starts moving backward along the template. During backtracking, the 3′ end of the nascent transcript disengages from the transcription bubble and enters the RNA polymerase secondary channel. Resumption of elongation can only occur after the disengaged 3′ end is cleaved from the transcript. Pol III is capable of catalyzing this cleavage reaction unassisted and this may be a normal part of the transcriptional termination process. Bacterial RNA polymerase and Pol II catalyze the cleavage reaction very inefficiently in the absence of accessory cleavage factors. These accessory factors include the bacterial factors GreA and GreB and the eukaryotic factor TFIIS (p. 33).

Barr body – *see* X chromosome inactivation.

basal machinery – the set of transcription factors sufficient for promoter specific transcription in vitro. In bacteria the basal machinery simply consists of the core RNA polymerase plus a σ factor. In archaea and eukaryotes, the σ factor is replaced by a set of general transcription factors (p. 4).

basic-helix-loop-helix (bHLH) domain – a domain found in many eukaryotic sequence specific transcription factors. It consists of a positively charged α-helical segment (the basic region) followed by two amphipathic α-helices and an intervening loop (the helix-loop-helix). The basic regions binds in the DNA major groove, while the helix-loop-helix mediates dimerization (p. 87).

basic-leucine zipper (bZip) domain – a domain found in many eukaryotic sequence-specific transcription factors. It consists of a single α-helix. The first third of this domain is the positively charged basic region that binds in the DNA major groove. The remainder of the domain consists of a leucine zipper, which is a type of amphipathic α-helix. The leucine zipper mediates dimerization (p. 87).

Bicoid – a homeodomain-containing transcription factor that plays an essential role in defining the anterior/posterior polarity of the *Drosophila* embryo and in initiating the process of segmentation (Box 7.2).

Brahma – a member of the Trithorax group and the catalytic subunit of an ATP-dependent chromatin remodeling complex (equivalent to the budding yeast SWI/SNF complex) (p. 157).

B-recognition element (BRE) – *see* promoter.

B-related factor (BRF) – a subunit of TFIIIB, one of the general transcription factors required for transcription of eukaryotic class III genes. BRF exhibits structural and functional homology to the class II general transcription factor TFIIB (p. 52).

bromodomain – a protein domain often found in coactivators including many chromatin remodeling factors. It mediates binding to acetyl-lysine residues, especially in histone tails (p. 125).

budding yeast – *see Saccharomyces cerevisiae.*

calico cat – a cat characterized by patches of black and orange fur. Calico cats are invariably female. The coat pattern results from random inactivation of one X chromosome in each cell of the early embryo followed by faithful epigenetic inheritance of the active and inactive X chromosomes throughout the remainder of development (p. 160).

capping enzyme – *see* 7-methyl-Gppp cap.

catabolite gene activator protein (CAP) – *see* cyclic AMP receptor protein (CRP).

centromere – the portion of each eukaryotic chromosome surrounding the site to which the mitotic spindle attaches. The centromere is largely devoid of genes and contains repetitive DNA sequences. It is almost invariably heterochromatic in nature (p. 107).

chromatin – the nucleoprotein complex into which eukaryotic DNA is packaged. This packaging compacts DNA by up to six orders of magnitude allowing the meter of DNA found in a typical mammalian cell to fit inside a nucleus that is only a few microns across. Chromatin is about half DNA and half protein, with most of the protein consisting of small basic polypeptides termed histones. The basic repeating unit of chromatin is the nucleosome. Eight histone polypeptide chains (the histone octamer) associate with about 150 bp of DNA to form the nucleosome core. The histone octamer includes two copies of each of the four core histones: histone H2A, histone H2B, histone H3, and histone H4. In addition to the nucleosome core, the nucleosome includes linker DNA and associated linker histones, such as histone H1. The total distance along the DNA from one nucleosome to the next is typically ~160–200 bp. This linear array of nucleosomes has an approximate diameter of 10 nm and is termed the 10 nm chromatin fiber. Further compaction of the 10 nm chromatin fiber yields 30 nm chromatin fibers as well as even thicker ~100 nm fibers sometimes termed chromonema fibers. These chromonema fibers can also undergo further levels of compaction perhaps by the formation of large loops attached to a protein scaffold (p. 102).

chromatin immunoprecipitation (ChIP) assay – a technique using antibodies against a protein of interest (or a post-translation protein modification of interest) to determine the distribution of the protein (or protein modification) along a chromosome (p. 109 and Box 5.1).

chromatin remodeling – changes in chromatin structure brought about by ATP-dependent chromatin remodeling complexes. These protein complexes function to slide nucleosomes along the DNA, to strip nucleosomes off the DNA, or to alter nucleosome structure. This remodeling is often important to allow the transcriptional machinery access to promoters or to permit passage of the ternary elongation complex (p. 103).

chromodomain – a protein domain found in many coregulatory factors that mediate changes in chromatin structure. This domain binds to methylated lysine side chains in histone tails. They are usually highly specific for a particular methyl-lysine residue (p. 125).

chromomema fiber – see chromatin.

cis-element – a segment of DNA that serves as a recognition element and binding site for a component of the transcriptional machinery. Cis-elements include promoters as well as binding sites for sequence-specific transcription factors (p. 7).

cis-regulatory module (CRM) – a region of DNA that consists of multiple closely linked binding sites for sequence-specific transcription factors. The factors that bind these sites work together via combinatorial control mechanisms to control the activity of a linked promoter.

They can be upstream or downstream of the core promoter and can often exert their effects over long distances (up to millions of basepairs). CRMs are also sometimes termed enhancers if their primary role is in gene activation or silencers if their primary role is in gene repression (p. 50).

class I gene – a gene transcribed by RNA polymerase I (p. 21).

class II gene – a gene transcribed by RNA polymerase II (p. 21).

class III gene – a gene transcribed by RNA polymerase III (p. 21).

cleavage and polyadenylation – an RNA processing event that generates the 3′ end of most mature mRNAs. Following the stop codon of most class II genes is a cis-element known as the polyA signal. Cleavage factors recognize this element and cleave the nascent transcript at this signal. PolyA polymerase then adds a string of several hundred adenylate residues to the 3′ end. The purpose of this polyA tail is to protect the 3′ end of the mRNA from exonucleolytic degradation and to facilitate transport of the message to the ribosome (p. 44).

closed complex – *see* transcriptional initiation.

clr genes – a set of fission yeast genes discovered because of their requirement for silencing of the silent mating type loci. *clr* stands for cryptic loci regulator. These genes are required for the formation of both facultative and constitutive heterochromatin (p. 139).

Clr4 – the product of the fission yeast *clr4* gene. It encodes a histone lysine methyltransferase that catalyzes the di- and trimethylation of histone H3 lysine 9 and thereby plays a critical role in heterochromatin formation. Most eukaryotes contain orthologs of Clr4 that are equally important in heterochromatin formation. The *Drosophila* ortholog of Clr4 is Su(var)3–9 (p. 142).

coactivator – regulatory factors required for activation that do not bind directly to the DNA. They are instead recruited to the DNA by interactions with DNA-bound activator proteins (p. 6).

combinatorial control – regulation of gene activity by combinations of two or more regulatory factors acting in concert to determine the transcriptional state of a gene. The two basic forms of combinatorial control are synergy and antagonism (p. 168).

constitutive heterochromatin – *see* heterochromatin.

cooperativity – a property of certain multistep processes. A multistep process is said to exhibit cooperativity if the earlier steps facilitate the later steps. For example, in cooperative binding of transcription factors to CRMs, the binding of one factor to the CRM increases the likelihood of a second factor binding to the CRM. In extreme cases of cooperative binding, neither factor will bind the CRM alone, but only in the presence of the other factor (Box 7.1).

coordinate bond – a type of covalent bond that forms between a metal ion and an electronegative atom such as oxygen or nitrogen, in which both of the electrons that comprise the bond are donated by the electronegative atom. They are weaker than conventional covalent bonds, but stronger than ionic bonds (p. 25).

core promoter – *see* promoter.

core RNA polymerase – *see* RNA polymerase.

coregulator – a coactivator or a corepressor (p. 6).

corepressor – regulatory factors required for repression that do not bind directly to the DNA. They are instead recruited to the DNA by interactions with DNA-bound repressor proteins (p. 6).

CREB binding protein (CBP) – a protein first characterized as a coactivator for the cyclic AMP response element binding protein (CREB). It has since been shown to function as a coactivator for many activator proteins. It has histone acetyltransferase activity and can also serve as an adaptor between activators and RNA polymerase II (p. 90).

C-terminal domain (CTD) – a C-terminal extension of the largest subunit of RNA polymerase II. It consists of multiple repeats of the seven amino acid sequence YSPTSPS. The CTD has many important roles in transcription. For example, it forms part of the docking site for the binding of the Mediator to core RNA polymerase II. In addition, Ser2 and Ser5 of the seven amino acid sequence are phosphorylated and dephosphorylated in a highly regulated manner during the transcription cycle. This phosphorylation serves to control ternary elongation complex processivity and to coordinate transcription with RNA processing (p. 21).

cyclic AMP (cAMP) – a form of adenosine monophosphate in which the phosphate group is esterified to both the 3′ and 5′ hydroxyl groups of the ribose moiety. It is an important second messenger in all organisms and is formed from ATP by the enzyme adenylate cyclase. In bacteria, cAMP forms in response to glucose deprivation and then binds and activates a transcription factor termed cyclic AMP receptor protein (CRP). In eukaryotes, cAMP forms in response to a wide variety of stimuli and then activates protein kinase A, which, in turn phosphorylates many substrates including a transcription factor termed the cyclic AMP response element binding protein (CREB) (p. 73).

cyclic AMP receptor protein (CRP) – a helix-turn-helix motif-containing bacterial activator. CRP binds adjacent to and activates the promoters of operons required for the metabolism of sugars other than glucose. Also known as catabolite gene activator protein (CAP) (p. 73).

cyclic AMP response element binding protein (CREB) – an activator protein. CREB is phosphorylated by protein kinase A at a single serine residue within an activation domain termed the kinase inducible domain (KID). Phosphorylation of KID leads to binding of KID to the coactivator CREB binding protein (CBP) (p. 74).

cyclin-dependent kinases (CDKs) – a family of protein kinases. These kinases are only active when bound to a member of the cyclin protein family. Several members of the CDK family help to regulate transcription, including Cdk7 (a TFIIH subunit), Cdk8 (a Mediator subunit), and Cdk9 (a P-TEFb subunit) (p. 67).

Cys_2His_2 zinc finger motif – a DNA binding domain found in many eukaryotic transcription factors. Its structure is stabilized by a zinc ion forming coordinate bonds to two Cys residues and two His residues (p. 88).

Cys_4Cys_4 domain – a DNA binding domain found in many eukaryotic transcription factors. It is characteristic of virtually all nuclear receptors. Its structure is stabilized by two zinc ions, each one forming coordinate bonds with four Cys residues (p. 88).

Dicer – *see* RNA interference.

DNA helicase – enzymes that couple energetically favorable ATP hydrolysis to the energetically unfavorable formation of single-stranded DNA from double-stranded DNA. At class II promoters, formation of the transcription bubble (open complex formation) requires the DNA helicase XPB, which is a subunit of the general transcription factor TFIIH (p. 65).

DNA methyltransferase – *see* 5-methylcytosine.

domains of life – the three major branches of the tree of life, including bacteria, archaea, and eukaryotes (Box 1.1).

dosage compensation – the processes that alter the level of expression of genes located on the X chromosome to compensate for the fact that in some species females contain two X chromosomes, while males contain one X chromosome. In *Drosophila*, dosage compensation occurs by the two-fold upregulation of the entire X chromosome in males. In mammals, dosage compensation occurs by the inactivation of one X chromosome in each cell of the female (p. 159).

downstream promoter element (DPE) – *see* promoter.

DRB sensitivity inducing factor (DSIF) – a eukaryotic elongation factor. Together with negative elongation factor (NELF), DSIF binds to RNA polymerase and induces pausing of Pol II. Release from the paused state is promoted by Positive Transcription Elongation Factor b (P-TEFb), which catalyzes phosphorylation of the Pol II CTD triggering the dissociation of NELF. The nucleotide analogue 5,6-dichloro-1-β-D-ribofuranosyl-benzimidazole (DRB), which is a P-TEFb inhibitor, only interferes with transcriptional elongation if DSIF is available to induce pausing (p. 96).

Drosophila melanogaster – a species of fruit fly – *Drosophila*, for short. This model organism has been used in a wide variety of genetic studies since early in the 20th century when experiments on *Drosophila* led to the rediscovery of Mendel's laws and to the discovery that genes are linked together on chromosomes. Many features of this organism, including its relatively simple and completely sequenced genome ($\sim 1.3 \times 10^8$ bp, ~14,000 genes arrayed on four chromosomes), its short generation time (~10 days), and its high fecundity (each female can produce ~500 offspring), make it particularly amenable to genetic analysis. Recent investigation of *Drosophila* development using combined genetic, biochemical, and molecular biological approaches have illuminated the genetic hierarchy that controls embryogenesis demonstrating the over-riding importance of transcriptional regulation in development (Box 7.2).

enhanceosome – a nucleoprotein complex that forms at enhancers. It includes the enhancer DNA, activators (including conventional activators and architectural factors), and coactivators. Assembly of the enhanceosome is cooperative, meaning that the absence of any one component will significantly destablize the complex. The enhanceosome provides a mechanistic explanation for synergy (p. 173).

enhancer – *see* cis-regulatory module (CRM).

enhancer of variegation (E(var)) – see *white^{m4}*.

Enhancer of zeste (E(z)) – a member of the Polycomb group and a histone methyltransferase that catalyzes the trimethylation of histone H3 lysine 27. E(z) is the catalytic subunit of Polycomb repressive complex 2 (PRC2) and contains a SET domain (p. 154).

epi-allele – *see* epigenetics.

epigenetics – the study of heritable variation that does not result from variation in the DNA sequence. Such variation often results from heritable changes in chromatin structure that lead to changes in transcriptional activity. These include changes in the histone post-translational modification state (especially histone methylation) or changes in cytosine methylation. These heritable chromatin structures are sometimes termed epigenetic states or epi-alleles. Epigenetic states are not only transmitted from cell generation to cell generation during mitosis, but can also be transmitted from parent to offspring, obeying the laws of Mendelian inheritance (p. 102 and Box 6.1).

equilibrium sliding model – a simple model to explain the directional movement (translocation) of RNA polymerase along the template during transcriptional elongation. According to this model, RNA polymerase can move back and forth along the template so that template nucleotide n (the nucleotide basepaired to the residue at the 3′ end of the growing RNA strand) and template nucleotide n+1 (the next nucleotide to be copied) alternately occupy the catalytic site. Energetically favorable phosphodiester bond formation, which can only occur when template nucleotide n+1 occupies the active site, then serves to drive the enzyme down the template in a directional manner (p. 33).

euchromatin – one of two major forms of chromatin found in eukaryotic cells, the other form being heterochromatin. During interphase, euchromatin consists largely of ~100 nm diameter fibers (sometimes termed chromonema fibers). During mitosis, however, the euchromatin condenses giving rise to the compact mitotic chromosomes. The vast majority of the active genes reside in euchromatin, which is the transcriptionally active portion of the genome. Note, however, that in a complex eukaryotic genome only a small percentage of the euchromatic genes are transcribed at any one time (p. 106).

even-skipped (*eve*) – a *Drosophila* segmentation gene of the pair rule class that encodes a homeodomain-containing transcription factor (box 7.2).

facultative heterochromatin – *see* heterochromatin.

fission yeast – *see Schizosaccharomyces pombe*.

Gal4 – a yeast activator containing a Zn_2Cys_6 binuclear cluster (p. 87).

Gal4-VP16 – a chimeric activator consisting of the Gal4 DNA binding domain fused to the VP16 acidic activation domain (p. 86).

Gcn4 – a yeast activator containing a bZip domain (p. 87).

Gcn5 – a yeast histone acetyltransferase and the catalytic subunit of the SAGA complex (p. 111).

general transcription factors – proteins employed by eukaryotic and archaeal RNA polymerases to recognize promoters and initiate transcription. Each eukaryotic and archaeal RNA polymerase utilizes a distinct set of general transcription factors, although there is some overlap between the sets (p. 36 and Table 3.1).

giant (*gt*) – a *Drosophila* segmentation gene. It encodes a transcription factor that functions as a short-range repressor to define the anterior border of *eve* stripe 2 (and Box 7.2).

GNAT family complexes – a family of histone acetyltransferase complexes (including the SAGA complex) that are all characterized by a catalytic subunit closely related to Gcn5 (Table 5.1).

heat shock factor (HSF) – an activator that mediates the response to stress. In response to heat shock or other stresses, HSF trimerizes and binds to heat shock elements (HSEs) in heat shock genes activating their expression (p. 95).

HeLa cell – a human cervical carcinoma cell line that is frequently used in studies of transcription (p. 58).

helix-turn-helix (HTH) motif – a structural motif found in many transcription factors that mediates sequence-specific binding to DNA. It consists of two α-helices separated by a turn. The two α-helices pack against one another at roughly right angles, with the second helix (the recognition helix) resting in the DNA major groove making sequence specific contacts with the major groove edges of the basepairs (p. 76).

heterochromatic silencing – the repression of gene expression associated with the formation of heterochromatin, including both facultative and constitutive heterochromatin. The transcriptionally silent state associated with heterochromatin is an epigenetic state that can be transmitted from cell generation to cell generation and from parent to offspring (p. 137).

heterochromatin – one of two major forms of chromatin in the eukaryotic cell, the other form being euchromatin. Unlike euchromatin, which condenses at the beginning of mitosis and then decondenses after the completion of mitosis, heterochromatin remains highly condensed throughout the cell cycle. It tends to associate with the nuclear envelope. With a few exceptions, heterochromatin is transcriptionally silent. Much of the heterochromatin is associated with centromeres and telomeres. Since these regions are almost always heterochromatic in normal nuclei, centromeric and telomeric heterochromatin are termed constitutive heterochromatin. A second form of heterochromatin, termed facultative heterochromatin, forms at discrete regions within the chromosome arms for the purpose of silencing gene expression (p. 106).

heterochromatin protein 1 (HP1) – the *Drosophila* and mammalian counterpart of the fission yeast protein Swi6. In *Drosophila*, HP1 is encoded by the gene *Su(var)2–5*. It is bound throughout the heterochromatic regions of the genome and stabilizes heterochromatin. Like Swi6, it contains a chromodomain that binds the di- and trimethylated form of histone H3 lysine 9 (p. 147).

heteroduplex – a nucleic acid duplex consisting of one strand of DNA and one strand of RNA. The transcription bubble associated with the RNA polymerase ternary elongation complex contains an 8–9 bp long heteroduplex in which the 3′ end of the nascent RNA is paired with the template strand of the DNA (p. 29).

histone – *see* chromatin.

histone acetyltransferase (HAT) – an enzyme that acetylates lysine residues in histone tails using acetyl-coenzyme A as the acetyl group donor. Histone acetyl transferases are often the critical catalytic components of coactivator complexes such as the SAGA complex (p. 109).

histone code hypothesis – the idea that the post-translational modification state of each gene constitutes a code that determines the transcriptional state of a gene. More specifically, the hypothesis posits that histone post-translational modifications are recognized and bound by regulatory factors that then alter the accessibility of the gene and/or the affinity of the gene for the transcriptional machinery (p. 102).

histone deacetylase (HDAC) – an enzyme that hydrolyzes the amide linkage in an acetylated lysine side chain in a histone tail, releasing acetate and regenerating the unmodified lysine residue. Histone deacetylases are often the critical catalytic components of corepressor complexes such as the Sin3 complex (p. 109).

histone fold domain (HFD) – a protein dimerization motif found in the core histones. It mediates the formation of H2A:H2B and H3:H4 dimers, which are components of the histone octamer. A dimer of two histone fold domains is termed a histone fold pair (p. 104).

histone lysine demethylase (HKDM) – an enzyme that removes methyl groups from histone methyl-lysine residues. Cleavage is achieved via oxidative mechanisms releasing the methyl group as formaldehyde and regenerating the unmodified lysine side chain. These enzymes exhibit high levels of specificity for particular histone lysine side chains (p. 118).

histone lysine methyltransferase (HKMT) – an enzyme that methylates lysine side chains in histone tails using S-adenosylmethionine as a methyl group donor. Most of these enzymes contain a catalytic SET domain and are exquisitely specific for a particular lysine side chain (p. 117).

histone tail – poorly ordered N-terminal and C-terminal extensions of the core histones. These protrude from the nucleosome core and mediate internucleosomal contacts. They are targets for a wide variety of post-translation modifications (p. 107).

HMG I – an architectural factor that binds to multiple sites in the IFNβ enhancer. By altering the curvature of the enhancer, HMG I renders binding of the other sequence-specific activators cooperative (p. 174).

homeodomain – a eukaryotic DNA binding domain. The homeodomain is found in a variety of factors with critical roles in embryonic development including the products of the homeotic genes. It contains a helix-turn-helix motif (p. 88).

homeotic gene complex – a group of genes encoding the homeodomain-containing transcription factors that determine segment identity in segmental organisms such as insects and vertebrates. Mutations in homeotic genes result in homeosis, which means the assumption by one segment of characteristics normally associated with another segment (p. 150).

hunchback – a *Drosophila* segmentation gene. It encodes a Cys_2His_2 zinc finger domain-containing transcription factor. Its many functions include activation of *eve* stripe 2 and repression of *Ubx* (Box 7.2).

initial transcribing complex – *see* transcriptional initiation.

initiator (Inr) – *see* promoter.

interferon response factor (IRF) – an activator protein that binds the interferon-β (IFNβ) enhancer (p. 173).

interferon-β (IFNβ) enhancer – a cis-regulatory module that activates expression of the IFNβ promoter by mediating the assembly of a nucleo-protein complex termed the IFNβ enhanceosome (p. 173).

intrinsic termination – *see* transcriptional termination.

K region – the 11 kb region in the fission yeast genome between the *mat2* and *mat3* loci. The fission yeast genome contains three mating type loci, *mat1*, *mat2*, and *mat3*. *mat1* is transcriptionally active and determines the mating type. *mat2* and *mat3* are transcriptionally silent due to silencers in the *K* region that direct heterochromatin formation (p. 139).

kinase A inducible activation domain (KID) – *see* cyclic AMP response element binding protein (CREB).

Kleinfelter's sydrome – a condition in mammals resulting from aberrant segregation of the sex chromosomes during meiosis. The afflicted individuals, who develop as males, contain two X chromosomes and one Y chromosome. X-chromosome inactivation silences one of the two X chromosomes (p. 160).

Kruppel (*Kr*) – a *Drosophila* segmentation gene. It encodes a Cys_2His_2 zinc finger domain-containing transcription factor, the many functions of which include repression of *eve* stripe 2 to determine the posterior border of this stripe (Box 7.2).

λcI protein – a helix-turn-helix motif-containing transcription factor encoded by the bacteriophage λ genome. It can function as either an activator or a repressor depending on binding site context. Also termed the λ repressor (p. 73).

leucine-rich motif – a short motif with the consensus sequence LXXLL (where L is leucine and X can be any amino acid) found in many coactivator proteins. It mediates binding of the coactivators to the liganded ligand binding domains (LBDs) of many nuclear receptors (p. 187).

ligand-binding domain (LBD) – *see* nuclear receptor.

LXXLL motif – *see* leucine-rich motif.

Mad – *see* Myc family.

major groove – one of two depressions between the sugar phosphate chains of double-stranded DNA, the other being the minor groove. The two edges of the basepairs are accessible through these grooves (p. 76).

Max – *see* Myc family.

Mediator – a 20–30 subunit coactivator complex that is required for optimal activation by many, perhaps most, eukaryotic activators. In addition to contacting activators through diverse surfaces, the Mediator makes extensive contact with core RNA polymerase II. In yeast, the Mediator was initially isolated in a complex with core Pol II termed the RNA polymerase holoenzyme. The Mediator stimulates transcription at multiple steps including RNA polymerase recruitment as well as steps following recruitment. Many of the subunits of the yeast Mediator were initially identified in a suppressor screen to find mutations that suppressed the cold-sensitive phenotype due to truncation of the RNA polymerase II C-terminal domain (p. 90).

5-methylcytosine – a modified form of cytosine found in DNA. It arises post-transcriptionally when enzymes termed DNA methyltransferases transfer a methyl group from S-adenosylmethionine to the 5-carbon of a cytosine ring in DNA. It is largely absent from yeast and fruit flies, but is relatively abundant in vertebrate and plant genomes. In these organisms it is often found associated with transcriptionally silent genomic regions, including heterochromatin (p. 148).

7-methyl-Gppp cap – the 7-methylguanosine-5′-triphosphate residue that is attached to the 5′ end of all eukaryotic mRNAs. It is attached to the first Pol II incorporated nucleoside residue via a 5′ to 5′ triphosphate linkage. The enzyme responsible for forming this linkage, termed the capping enzyme, binds the RNA polymerase II C-terminal domain putting it in position to catalyze cap formation as soon as the 5′ end of the transcript emerges from the RNA exit channel. The purpose of this cap is to block degradation of the mRNA by 5′ exonucleases and to serve as a feature for recognition by the translational machinery (p. 68).

microcin J25 – a 21 amino acid residue-long peptide lariat that functions as a potent inhibitor of bacterial RNA polymerases. It binds to the secondary channel preventing nucleotides from reaching the active site (p. 31).

minor groove – *see* major groove

mitogen activated protein (MAP) kinase – a serine/threonine protein kinase found close to the end of a protein phosphorylation cascade triggered by receptor tyrosine kinases. MAP kinase directly phosphorylates numerous transcription factors altering their activity and therefore the transcriptional program of the cell (p. 189).

Myc family – a family of basic-helix-loop-helix domain-containing transcription factors including Myc, Mad, and Max. The Myc/Max heterodimer generally functions as an activator by recruiting histone acetyltransferase complexes, while the Mad/Max heterodimer generally functions as a repressor by recruiting histone deacetylase complexes (p. 114).

MYST complexes – a family of histone acetyltransferase complexes that are characterized by catalytic subunits with homology to Esa1 (p. 116).

nascent RNA – literally means an RNA molecule being born. The transcript found in the ternary elongation complex is a nascent RNA (p. 41).

NCoR – a corepressor that binds to the unliganded ligand binding domains of many nuclear receptors (p. 188).

negative elongation factor (NELF) – *see* DRB sensitivity inducing factor (DSIF).

NF-κB – an activator protein that is a component of the IFNβ enhanceosome (p. 173).

NOT operator – *see* antagonism.

nuclear receptors – as large family of sequence-specific transcription factors that contain Cys_4Cys_4 DNA binding domains. Most nuclear receptors are regulated by ligands, which bind to their ligand binding domains (LBDs) and modulate their activity. The two major classes of nuclear receptors are the class I receptors, which bind steroid hormones such as estrogens and glucocorticoids, and the class II receptors, which bind non-steroidal lipophilic ligands such as fatty acids, retinoids, thyroid hormone, and vitamin D. Members of the family lacking known ligands are termed orphan receptors (p. 185).

nucleosome – *see* chromatin.

nucleosome remodeling and deacetylation (NuRD) complex – a coregulatory complex that contains both histone deacetylase and nucleosome remodeling activity (p. 113).

nucleotide excision repair – a form of DNA repair in which a stretch of damaged DNA is excised by the combined action of DNA helicases and nucleases. The resulting gap is then filled in and sealed by the combined action of DNA polymerases and DNA ligases. Two of the DNA helicases required for nucleotide excision repair, XPB and XPD, are components of the class II general transcription factor TFIIH. The involvement of these factors in both DNA repair and transcription may provide an explanation for transcription coupled repair in which actively transcribed regions of the genome are more efficiently repaired than transcriptionally silent regions (p. 66).

open complex – *see* transcriptional initiation.

pair-rule genes – a group of segmentation genes essential for the subdivision of the *Drosophila* embryo into a series of repeating segments. These genes, including *even-skipped (eve)*, are often expressed in a seven stripe pattern along the anterior-posterior axis of the early embryo (Box 7.2).

Pcf11 – *see* transcriptional termination.

Pcr1 – *see* Atf1.

Pnt-P2 – a *Drosophila* activator containing an ets DNA binding domain. Receptor tyrosine kinase (RTK) signaling leads to the phosphorylation and consequent activation of Pnt-P2 (p. 189).

Polycomb – *see* PRC1.

Polycomb group (PcG) – one of two groups of genes whose products are required for the epigenetic maintenance of homeotic gene expression throughout development, the other group being the Trithorax group. Early during embryogenesis, transcription factors encoded by the segmentation genes set up the spatially regulated patterns of homeotic gene expression. Once these factors disappear from the embryo after the first ~12 hours of embryogenesis, the products of the Polycomb and Trithorax group genes ensure that the spatially restricted homeotic gene expression patterns will be maintained for the remaining 10 days of the developmental cycle. The Polycomb group proteins maintain the transcriptionally silent state, while the Trithorax group proteins maintain the transcriptionally active state (p. 153).

Polycomb repressive complex 1 (PRC1) – a protein complex composed of products of several of the Polycomb group genes. One subunit of this complex, Polycomb, contains a chromodomain that recognizes trimethylated histone H3 lysine 27, a post-translational modification intimately associated with Polycomb group-mediated repression (p. 154).

Polycomb repressive complex 2 (PRC2) – a histone methyltransferase complex composed of products of several of the Polycomb group genes. The catalytic subunit of this complex is the SET domain-containing factor Enhancer of zeste. PRC2 catalyzes the trimethylation of histone H3 lysine 27, a post-translational modification intimately associated with Polycomb group-mediated repression (p. 154).

Polycomb response element (PRE) – cis-regulatory elements in the *Drosophila* homeotic gene complex required for maintenance of the silent state by the Polycomb group. PREs often contain binding sites for the Polycomb group gene product Pleiohomeotic (Pho), which binds these sites and helps to recruit other Polycomb group proteins (p. 156).

position effect variegation (PEV) – *see white^{m4}*.

positive control (pc) mutation – a mutation in an activator protein that eliminates activation, but not other biochemical activities of the activator such as DNA binding and dimerization. Such mutations often define surfaces on activators that contact RNA polymerase or other components of the transcriptional machinery (p. 77).

positive transcription elongation factor b (P-TEFb) – *see* DRB sensitivity inducing factor (DSIF).

preinitiation complex (PIC) – a eukaryotic RNA polymerase and its complete set of general transcription factors bound to a core promoter. The preinitiation complex is equivalent to the bacterial closed promoter complex (p. 52).

primary channel – the channel between the two claws of a multisubunit RNA polymerase through which the DNA template enters the active site (p. 31).

processivity – a property of a polymerase such as an RNA polymerase. An RNA polymerase is said to be processive if it adds one nucleotide residue after another to the 3' end of a nascent RNA strand without releasing the nascent transcript or the template between elongation cycles (p. 32).

promoter – the DNA region around the transcriptional start site required for recognition and initiation by the general transcriptional machinery. In bacterial genes, the promoter generally consists of two conserved upstream elements, the –10 element (or Pribnow box) and the –35 element. In eukaryotic class II genes, promoters can be defined by numerous elements, including the TATA box, the initiator (Inr), and the downstream promoter element (DPE). Most class II promoters contain only one or two of these elements. The eukaryotic class II promoter is sometimes termed the core promoter. It is defined operationally as the region that is sufficient for transcriptional initiation in vitro by purified RNA polymerase II and the class II general transcription factors (p. 2).

promoter clearance – *see* transcriptional initiation.

promoter escape – *see* transcriptional initiation.

protein kinase A – *see* cyclic AMP (cAMP).

receptor tyrosine kinase (RTK) – a transmembrane protein that mediates the response to numerous extracellular signaling agents, usually peptides such as epidermal growth factor, fibroblast growth factor, and nerve growth factor. Binding of one of these ligands to its cognate RTK stimulates an intracellular protein kinase cascade leading to the phosphorylation and activation of mitogen activated protein (MAP) kinase, which, in turn, phosphorylates multiple transcription factors altering their properties (p. 188).

recognition helix – an α-helix in a DNA binding domain that is directly responsible for recognizing and binding a cis-regulatory element in a sequence specific manner. Recognition helices generally bind DNA in the major groove (p. 76).

recruitment – a mechanism for regulating transcription in which one factor (e.g., an activator) makes energetically favorable contacts with another factor (e.g., RNA polymerase or a coactivator) stabilizing the association of the second factor with the template (p. 85).

repressor – *see* sequence-specific transcription factor.

rifampicin – an inhibitor of bacterial RNA polymerases and an important antibiotic in the treatment of tuberculosis. Rifampicin binds in the active site of the core polymerase and blocks elongation when the transcript is only a few nucleotides in length (p. 31).

RNA exit channel – a channel in RNA polymerase through which the nascent transcript exits the active site. In bacterial core polymerase, the exit channel runs underneath a lobe of the β subunit termed the β-flap (p. 31).

RNA interference (RNAi) – the RNA guided silencing of gene expression. This regulation can occur at the level of transcription (as in heterochromatic silencing) or at steps in gene expression subsequent to transcription (post-transcriptional gene silencing [PTGS]). All arms of the pathway begin with double-stranded RNA, which is processed by an endonuclease termed dicer into 21–27 bp long fragments termed small interfering RNA (siRNA). These siRNAs then guide protein complexes containing the Argonaut protein to specific genomic regions or transcripts leading to gene-specific silencing of gene expression (p. 144). Argonaut-containing complexes include the RNA-induced initiation of transcriptional gene silencing (RITS) complex, which mediates transcriptional gene silencing, and RNA-induced silencing complex (RISC), which mediates PTGS (p. 144).

RNA polymerase – an enzyme that catalyzes the polymerization of nucleotide monophosphate residues using nucleotide triphosphate substrates. The multisubunit RNA polymerases use DNA as a template and are the major engines of transcription in all three domains of life. In bacteria, the RNA polymerase holoenzyme consists of the core polymerase, which contains five subunits and catalyzes transcriptional elongation, as well as a dissociable σ factor, which is required for promoter recognition. Eukaryotic and archaeal multisubunit RNA polymerase contain homologues of the five bacterial core enzyme subunits in addition to numerous additional subunits. These eukaryotic and archaeal enzymes are the equivalent of the core bacterial polymerase as they are unable to recognize promoters on their own. However, eukaryotes and archaea lack σ factors and instead use sets of general transcription factors to assist in promoter recognition and transcriptional initiation (p. 2).

Rpd3 – a yeast histone deacetylase and the catalytic subunit of the Sin3 and NuRD histone deacetylase complexes. Mammalian HDAC1 and HDAC2 are closely related to Rpd3 (p. 111).

Saccharomyces cerevisiae – budding yeast. This single cell eukaryote is frequently used as a model organism in studies of transcription. Its small and well-characterized genome greatly facilitate genetic analysis (p. 149).

SAGA complex – a GNAT family histone acetyltransferase complex in budding yeast. Its catalytic subunit is Gcn5. The SAGA complex contains products of many of the SPT (suppressor of Ty) genes and of the ADA (adaptor) genes, and SAGA is an acronym for Spt-Ada-Gcn5 acetyltransferase. The human counterpart of the SAGA complex is termed the STAGA complex (p. 111).

Schizosaccharomyces pombe – fission yeast. In many respects, the regulatory mechanisms used in metazoans more closely resemble the mechanisms in fission yeast than in budding yeast. For example, the heterochromatin forming machinery including the RNAi machinery in metazoans is present in fission yeast, but not in budding yeast (p. 137).

secondary channel – a pore in multisubunit RNA polymerases through which nucleotides are thought to reach the active site during transcriptional elongation (p. 31).

segmentation gene – a gene that directs the subdivision of the *Drosophila* embryo into a series of repeating segments (or metameres). Many of these genes encode transcription factors that work together in a regulatory hierarchy to define the segments. These transcription factors are also responsible for the establishment of the spatially regulated patterns of homeotic gene expression during early embryogenesis (Box 7.2).

sequence-specific transcription factor – a transcription factor that binds directly to cis-regulatory elements in the DNA in a sequence specific manner to control transcription. Sequence-specific transcription factors are also termed activators if they serve to increase rates of transcription or repressors if they serve to decrease rates of transcription (p. 75).

SET domain – a protein domain with histone lysine methyltransferase activity. SET is an acronym for three SET-domain containing *Drosophila* gene products – Su(var)3–9, Enhancer of zeste, and Trithorax (p. 137).

short interfering RNA (siRNA) – *see* RNA interference.

σ factor – *see* RNA polymerase.

silencer – *see* cis-regulatory module (CRM).

Sin3 complex – a histone deacetylase complex that contains Sin3, Rpd3 (the catalytic subunit), and several additional subunits. The Sin3 subunit often mediates interactions of this complex with repressors (p. 113).

SIR complex – a complex of three proteins that is required for heterochromatic silencing in budding yeast (*S. cerevisiae*). This complex, which is unique to budding yeast, includes the Sir2, Sir3, and Sir4 proteins. Sir2, a NAD$^+$-dependent histone deacetylase is the only member of this complex found in other eukaryotes (p. 149).

Sp1 – a Cys$_2$His$_2$ zinc finger domain-containing mammalian transcription factor (p. 88).

STAGA complex – *see* SAGA complex.

stripe enhancer – an enhancer found in a pair rule gene that directs one or sometimes two transverse stripes of gene expression in the early *Drosophila* embryo (p. 180).

Su(var)3–9 – *see* Clr4.

suppressor of variegation (Su(var)) – see *white*^{m4}.

SWI/SNF complex – an ATP-dependent chromatin remodeling complex. The catalytic subunit of this complex is Snf2 (p. 121).

Swi6 – *see* heterochromatin protein 1 (HP1).

synergy – a type of combinatorial control in which two or more transcriptional activators or coactivators working together yield levels of transcription that are greater than the sum of the activities observed for the factors working alone. In the most extreme form of synergy, individual factors fail to activate at all on their own and activation is only observed in the presence of two or more factors. In mathematical terms, this extreme form of synergy can be described with the use of the AND operator since gene activity requires the presence of activators X AND Y (p. 169).

TATA binding protein (TBP) – a "universal" transcription factor required for all transcription in eukaryotes and archaea. It has a very unusual saddle-shaped ten-stranded β-sheet that binds in the minor groove of the TATA box. This binding results in severe distortion of the DNA. In eukaryotes, it associates with several different sets of TBP associated factors (TAFs) to form complexes such as SL1 (a class I general transcription factor), TFIID (a class II general transcription factor), and TFIIIB (a class III general transcription factor) (p. 52).

TATA box – *see* promoter.

TBP associated factor (TAF) – *see* TATA binding protein (TBP).

telomeres – regions of repetitive DNA found at the ends of linear eukaryotic chromosomes. Telomeres are almost always heterochromatic (p. 107).

ternary elongation complex (TEC) – *see* transcriptional elongation.

thermophilic bacteria – bacteria, such as *Thermus aquaticus*, adapted to grow at high temperatures. The proteins from these organisms are of high stability to permit function at high temperature. As a result, they often yield high-quality crystals and are therefore favored for structure determination by X-ray crystallographers (p. 19).

transcriptional elongation – the phase of the transcription cycle during which a core RNA polymerase moves down the template adding nucleotide residues to the 3′ end of the nascent RNA. Each cycle of nucleotide addition includes a chemical catalysis phase in which a new phosphoester linkage is formed as well as a translocation phase in which the polymerase moves one basepair down the template placing the new 3′ end of the transcript in the catalytic site in preparation for the next round of catalysis. Transcriptional elongation is usually very processive – the polymerase does not generally release the transcript until a termination signal is encountered. The stable complex of core RNA polymerase, the DNA template, and the nascent transcript is termed the ternary elongation complex (TEC) (p. 23).

transcriptional initiation – the phase of the transcription cycle during which the promoter is recognized by RNA polymerase and RNA synthesis begins. This phase can be subdivided into multiple subphases, including closed complex formation (the binding of RNA polymerase to the promoter), open complex formation (the opening of the DNA around the transcriptional start site to form the transcription bubble), initial transcribing complex (also termed abortive initiation, during which RNA polymerase remains in contact with the promoter and repeatedly synthesizes and releases short transcripts), and promoter clearance (during which RNA polymerase releases the promoter leading to the formation of the highly processive ternary elongation complex) (p. 36).

transcriptional termination – the final step in transcription during which the ternary elongation complex falls apart releasing the template and the transcript from RNA polymerase. In bacteria, roughly half of the genes terminate via the intrinsic termination pathway, which does not rely on any accessory factors. Transcriptional termination at eukaryotic class I and class III genes occurs by a mechanism that is partially reminiscent of the bacterial intrinsic termination pathway. In contrast, termination at eukaryotic class II genes is more complex, requiring the action of factors such as Pcf11, which coordinate termination with cleavage and polyadenylation (p. 42).

transcription-coupled DNA repair – *see* nucleotide excision repair (NER).

transition state – an unstable high energy state through which a chemical reaction proceeds as reactants are converted to products. The difference between the free energy of the reactants and the free energy of the transition state is directly related to the rate of a reaction. Catalysts such as enzymes generally accelerate reactions by reducing this difference (p. 28).

translocate – to move along a polymer usually in a linear fashion. RNA polymerases, for example, are said to translocate down the DNA template during transcriptional elongation (p. 29).

Trithorax (Trx) – a member of the Trithorax group and a histone lysine methyltransferase that trimethylates histone H3 lysine 4 (p. 157).

Trithorax group (TrxG) – *see* Polycomb group (PcG).

TRRAP – a large subunit shared by many histone acetyltransferase (HAT) complexes including SAGA and MYST family complexes. Activator proteins often contact HAT complexes via this subunit. The yeast equivalent of TRRAP is termed Tra1 (p. 114).

ubiquitin – a small polypeptide that becomes covalently conjugated to other proteins via an amide linkage between the C-terminus of ubiquitin and the amino group of the lysine side chain in the target protein. Conjugation of ubiquitin to proteins (ubiquitylation) is most often a means of marking them for degradation. However, in some cases ubiquitylation leads to consequences other than degradation. For example, histone ubiquitylation is a means of modifying the activity of associated genes (p. 119).

Ultrabithorax (Ubx) – a *Drosophila* homeodomain-containing transcription factor encoded by one of the genes of the homeotic gene complex (p. 151).

VP16 – a coactivator encoded by herpes simplex virus. VP16 contains an extremely potent acidic activation domain and has been used extensively in studies of transcriptional activation. In a widely studied chimeric transcription factor termed Gal4-VP16, the DNA binding domain of the yeast activator Gal4 is fused to the VP16 activation domain (p. 86).

white^{m4} – a mutant allele of the *Drosophila white* gene. The *m4* allele designation stands for *mottled 4*. This gene, which resides on the X chromosome, is required for the normal red pigmentation of the compound fly eye and thus complete loss of the *white* gene leads to pure white eyes. In *white^{m4}* flies, a large chromosomal inversion has placed *white* close to centromeric heterochromatin. Spreading of heterochromatin from the centromere leads to silencing of *white*. Because the extent of spreading is variable, the flies exhibit variegated eyes containing patches of pigmented tissue and patches of unpigmented tissue. This variable inactivation of a gene due to aberrant positioning near a heterochromatic region of the genome is termed position effect variegation (PEV). Mutations that increase spreading and/or stability of heterochromatin leading to more unpigmented eye tissue define *enhancer of variegation (E(var))* genes, while mutations that decrease spreading and/or stability of heterochromatin leading to less unpigmented eye tissue define *suppressor of variegation (Su(var))* genes (p. 147).

X chromosome inactivation – the mechanism used to achieve dosage compensation in placental mammals. In this process, all but one X chromosome in each somatic cell of the early embryo is converted into transcriptionally silent heterochromatin. The resulting densely staining heterochromatic X chromosomes are termed Barr bodies. Thus, each cell in the normal female contains one inactive Barr body and one active X chromosome. The X chromosome in each cell to be inactivated is selected randomly, but once a chromosome is selected for inactivation it remains inactive throughout the remaining cell divisions leading to the mature adult. As a result, the adult female is a mixture of two types of cells, one type in which the paternal X chromosome is inactive and another type in which the maternal X chromosome is inactive (p. 159).

X-inactivation center (XIC) – the region on the mammalian X chromosome that directs X chromosome inactivation. It contains a gene termed Xist that gives rise to an RNA product that does not encode any protein. The Xist RNA spreads along the X chromosome from the XIC and recruits factors that direct heterochromatic silencing (p. 160).

Yan – a *Drosophila* repressor that is a substrate for mitogen activated protein (MAP) kinase. Phosphorylation of Yan by MAP kinase in response to receptor tyrosine kinase (RTK) signaling results in the nuclear export of Yan and therefore in the loss of repression by this factor (p. 189).

Zn_2Cys_6 binuclear cluster – a DNA binding domain found in some fungal transcription factors such as Gal4. Its fold is stabilized by two zinc ions, coordinating a total of six Cys side chains. Two of the Cys side chains are coordinated by both zinc ions and thus each zinc ion has a total of four Cys ligands (p. 89).

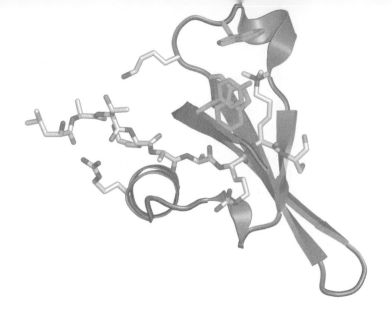

Index

Page numbers in *italic* refer to figures and/or tables separate from the text. Page numbers in **bold** refer to text and figures in boxes. Numbers preceded by P (e.g., P4.2) refer to plates